Memoirs
of the
American Mathematical Society

ISSN 0065-9266 (print)
ISSN 1947-6221 (online)

Number 1382

Hypergeometric Functions Over Finite Fields

Jenny Fuselier
Ling Long
Ravi Ramakrishna
Holly Swisher
Fang-Ting Tu

November 2022 • Volume 280 • Number 1382 (fifth of 8 numbers)

Library of Congress Cataloging-in-Publication Data

Cataloging-in-Publication Data has been applied for by the AMS.
See http://www.loc.gov/publish/cip/.
DOI: https://doi.org/10.1090/memo/1382

Memoirs of the American Mathematical Society

This journal is devoted entirely to research in pure and applied mathematics.

Subscription information. Beginning with the January 2010 issue, *Memoirs* is accessible from www.ams.org/journals. The 2022 subscription begins with volume 275 and consists of six mailings, each containing one or more numbers. Subscription prices for 2022 are as follows: for paper delivery, US$1085 list, US$868 institutional member; for electronic delivery, US$955 list, US$764 institutional member. Upon request, subscribers to paper delivery of this journal are also entitled to receive electronic delivery. If ordering the paper version, add US$22 for delivery within the United States; US$85 for outside the United States. Subscription renewals are subject to late fees. See www.ams.org/help-faq for more journal subscription information. Each number may be ordered separately; *please specify number* when ordering an individual number.

Back number information. For back issues see www.ams.org/backvols.

Subscriptions and orders should be addressed to the American Mathematical Society, P. O. Box 845904, Boston, MA 02284-5904 USA. *All orders must be accompanied by payment.* Other correspondence should be addressed to 201 Charles Street, Providence, RI 02904-2213 USA.

Copying and reprinting. Individual readers of this publication, and nonprofit libraries acting for them, are permitted to make fair use of the material, such as to copy select pages for use in teaching or research. Permission is granted to quote brief passages from this publication in reviews, provided the customary acknowledgment of the source is given.

Republication, systematic copying, or multiple reproduction of any material in this publication is permitted only under license from the American Mathematical Society. Requests for permission to reuse portions of AMS publication content are handled by the Copyright Clearance Center. For more information, please visit www.ams.org/publications/pubpermissions.

Send requests for translation rights and licensed reprints to reprint-permission@ams.org.

Excluded from these provisions is material for which the author holds copyright. In such cases, requests for permission to reuse or reprint material should be addressed directly to the author(s). Copyright ownership is indicated on the copyright page, or on the lower right-hand corner of the first page of each article within proceedings volumes.

Memoirs of the American Mathematical Society (ISSN 0065-9266 (print); 1947-6221 (online)) is published bimonthly (each volume consisting usually of more than one number) by the American Mathematical Society at 201 Charles Street, Providence, RI 02904-2213 USA. Periodicals postage paid at Providence, RI. Postmaster: Send address changes to Memoirs, American Mathematical Society, 201 Charles Street, Providence, RI 02904-2213 USA.

© 2022 by the American Mathematical Society. All rights reserved.
This publication is indexed in *Mathematical Reviews*®, *Zentralblatt MATH*, *Science Citation Index*®, *Science Citation Index*™-*Expanded*, *ISI Alerting Services*^{SM}, *SciSearch*®, *Research Alert*®, *CompuMath Citation Index*®, *Current Contents*®/*Physical, Chemical & Earth Sciences*.
This publication is archived in *Portico* and *CLOCKSS*.
Printed in the United States of America.

∞ The paper used in this book is acid-free and falls within the guidelines established to ensure permanence and durability.
Visit the AMS home page at https://www.ams.org/

10 9 8 7 6 5 4 3 2 1   27 26 25 24 23 22

Contents

Acknowledgments   vii

Chapter 1. Introduction   1
   1.1. Overview   1
   1.2. Organization and the main results   2

Chapter 2. Preliminaries for the Complex and Finite Field Settings   7
   2.1. Gamma and beta functions   7
   2.2. Gauss and Jacobi sums   10
   2.3. Lagrange inversion   13
   2.4. A dictionary between the complex and finite field settings   14

Chapter 3. Classical Hypergeometric Functions   15
   3.1. Classical development   15
   3.2. Some properties of hypergeometric functions with $n = 1$ 16

Chapter 4. Finite Field Analogues   27
   4.1. Periods in the finite field setting   27
   4.2. Hypergeometric varieties   28
   4.3. Hypergeometric functions over finite fields   29
   4.4. Comparison with other finite field hypergeometric functions   30

Chapter 5. Some Related Topics on Galois Representations   33
   5.1. Absolute Galois groups and Galois representations   33
   5.2. Grössencharacters in the sense of Hecke   35
   5.3. Notation for the Nth power residue symbol   36
   5.4. Jacobi sums and Grössencharacters   38

Chapter 6. Galois Representation Interpretation   41
   6.1. Galois interpretation for $_1\mathbb{P}_0$ 41
   6.2. Generalized Legendre curves and their Jacobians   42
   6.3. Galois interpretation for $_2\mathbb{P}_1$ 46
   6.4. Some special cases of $_2\mathbb{P}_1$-functions   49
   6.5. Galois interpretation for $_{n+1}\mathbb{F}_n$ 50
   6.6. Zeta functions and hypergeometric functions over finite fields   51
   6.7. Summary   53

Chapter 7. A finite field Clausen formula and an application   55
   7.1. A finite field version of the Clausen formula by Evans and Greene   55
   7.2. Analogues of Ramanujan type formulas for $1/\pi$ 57

Chapter 8. Translation of Some Classical Results   61

8.1.	Kummer's 24 Relations	61
8.2.	A Pfaff-Saalschütz evaluation formula	65
8.3.	A few analogues of algebraic hypergeometric formulas	66

Chapter 9. Quadratic or Higher Transformation Formulas — 73
- 9.1. Some results related to elliptic curves — 73
- 9.2. A Kummer quadratic transformation formula — 75
- 9.3. The quadratic formula in connection with the Kummer relations — 83
- 9.4. A finite field analogue of a theorem of Andrews and Stanton — 88
- 9.5. Another application of Bailey cubic transformations — 95
- 9.6. Another cubic $_2\mathbb{F}_1$ formula and a corollary — 97

Chapter 10. An application to Hypergeometric Abelian Varieties — 105

Chapter 11. Open Questions and Concluding Remarks — 109
- 11.1. Numeric observations — 109

Chapter 12. Appendix — 113
- 12.1. Bailey $_3F_2$ cubic transforms — 113
- 12.2. A proof of a formula by Gessel and Stanton — 114

Bibliography — 117

Index — 123

Abstract

Building on the developments of many people including Evans, Greene, Katz, McCarthy, Ono, Roberts, and Rodriguez-Villegas, we consider period functions for hypergeometric type algebraic varieties over finite fields and consequently study hypergeometric functions over finite fields in a manner that is parallel to that of the classical hypergeometric functions. Using a comparison between the classical gamma function and its finite field analogue the Gauss sum, we give a systematic way to obtain certain types of hypergeometric transformation and evaluation formulas over finite fields and interpret them geometrically using a Galois representation perspective. As an application, we obtain a few finite field analogues of algebraic hypergeometric identities, quadratic and higher transformation formulas, and evaluation formulas. We further apply these finite field formulas to compute the number of rational points of certain hypergeometric varieties.

Received by the editor April 5, 2016, and, in revised form, December 18, 2017, and July 16, 2019.

Article electronically published on October 7, 2022.

DOI: https://doi.org/10.1090/memo/1382

2020 *Mathematics Subject Classification.* 11T23, 11T24, 33C05, 33C20, 11F80, 11S40.

Key words and phrases. Hypergeometric functions, finite fields, Galois representations.

The first author is affiliated with High Point University, High Point, NC 27268, USA. Email: jfuselie@highpoint.edu.

The second author is affiliated with Louisiana State University, Baton Rouge, LA 70803, USA. Email: llong@lsu.edu.

The third author is affiliated with Cornell University, Ithaca, NY 14853, USA. Email: ravi@math.cornell.edu.

The fourth author is affiliated with Oregon State University, Corvallis, OR 97331, USA. Email: swisherh@math.oregonstate.edu.

The fifth author is affiliated with Louisiana State University, Baton Rouge, LA 70803, USA. Email: ftu@lsu.edu.

©2022 American Mathematical Society

Acknowledgments

Both Long and Tu were supported by NSF DMS #1303292 and Long was further supported by NSF DMS #1602047. Ramakrishna was supported by Simons Foundation Grant #524863. We would like to thank Microsoft Research and the Office of Research & Economic Development at Louisiana State University (LSU) for support which allowed Fuselier and Swisher to visit LSU in 2015. Tu was supported in part by the National Center for Theoretical Sciences (NCTS) in Taiwan for her visit to LSU in 2015. We are very grateful to George Andrews, Frits Beukers, Irene Bouw, Luca Candelori, Henri Cohen, Henri Darmon, Ron Evans, Sharon Frechette, Jerome Hoffman, Wen-Ching Winnie Li, Matt Papanikolas, Dennis Stanton, John Voight, Daqing Wan, Tonghai Yang and Wadim Zudilin for their interest, enlightening discussions/suggestions and/or helpful references. Special thanks go to Dennis Stanton who gave us two interesting problems and to Yifan Yang for sharing his results which are stated in Example 3.7. Special thanks also go to Rodriguez-Villegas for his lecture on Hypergeometric Motives given at NCTS in 2014. Our statements are verified by Magma or Sage programs. The collaboration was carried out on SageMathCloud. Long, Swisher and Tu further thank the Institute for Computational and Experimental Research in Mathematics (ICERM) at Brown University for its hospitality as part of this work was done while they were ICERM research fellows during Fall 2015. Finally, we thank the referees for the extremely careful reading of our work and the many detailed and helpful suggestions.

CHAPTER 1

Introduction

1.1. Overview

Starting with seminal works of Evans [**29, 31, 32**], Greene [**41**–**43**], Katz [**47**], Koblitz [**49**], McCarthy [**70, 71**], Ono [**75**], Roberts and Rodriguez-Villegas [**80**], Greene and Stanton [**44**] et al., the finite field analogues of special functions, in particular generalized hypergeometric functions, have been developed theoretically, and are known to be related to various arithmetic objects ([**16**], [**35**], [**37**], [**61**], [**72**] et al.). Computations on the corresponding hypergeometric motives have been implemented in computational packages like Pari and Magma, in particular, see [**99**] for the Magma documentation by Watkins.

We focus on the finite field analogues of classical transformation and evaluation formulas based on the papers mentioned above, especially [**41**] by Greene and [**70**] by McCarthy. To achieve our goals, we modify their notation slightly so that the analysis more closely parallels the classical case, and the proofs in the complex case can be extended naturally to the finite field setting. We emphasize that the classical $_{n+1}F_n$ functions with rational parameters and suitable normalizing factors can be computed from the $_{n+1}P_n$ period functions (see (3.6)). These are related to the periods of the hypergeometric varieties of the form

$$(1.1) \qquad y^N = x_1^{i_1} \cdots x_n^{i_n} (1-x_1)^{j_1} \cdots (1-x_n)^{j_n} (1 - \lambda x_1 x_2 \cdots x_n)^k,$$

where N, i_s, j_t, k are positive integers, λ is a fixed parameter, and y, x_m are variables. It is helpful to distinguish the $_{n+1}F_n$ functions from the period functions $_{n+1}P_n$. In particular we consider their finite field analogues, which we denote by $_{n+1}\mathbb{F}_n$ and $_{n+1}\mathbb{P}_n$, respectively (see definitions (4.4) and (4.9)).

In the finite field setting, we use two main methods throughout. The first method is calculus-style and is based on the conversion between the classical and finite field settings that is well-known to experts. This is summarized in a table in §2.4. In the classical setting, the gamma function satisfies reflection and multiplication formulas (see Theorems 2.4 and 2.6) which yield many interesting hypergeometric identities. Roughly speaking, Gauss sums are finite field analogues of the gamma function [**32**] and they satisfy very similar reflection and multiplication formulas (see (2.10) and Theorem 2.7). The technicality in converting identities from the classical setting lies in analyzing the 'error terms' associated with Gauss sum computations (which are delta terms here, following Greene's work [**41**]). This approach is particularly handy for translating to the finite field setting a classical transformation formula that satisfies the following condition:

> (∗) it can be proved using only the binomial theorem, the reflection and multiplication formulas, or their corollaries (such as the Pfaff-Saalschütz formula (3.15)).

1

For the second method, we rely on the built-in symmetries for the $_{n+1}\mathbb{F}_n$ hypergeometric functions over finite fields, such as the Kummer relations for $_2\mathbb{F}_1$ (§3.2.3), which are very helpful for many purposes including getting around the error term analysis.

We emphasize alignment with geometry by putting the finite field analogues in the explicit context of hypergeometric varieties and their corresponding Galois representations. This gives us important guidelines as the finite field analogues, in particular evaluation formulas such as the Pfaff-Saalschütz identity, do not look exactly like the classical cases. From a different perspective, hypergeometric functions over finite fields are twisted exponential sums in the sense of [1, 2] and are related to hypergeometric motives [16, 47, 80], abstracted from cohomology groups of the algebraic varieties. This direction is pursued by Katz [47] and further developed and implemented by many others including Roberts and Rodriguez-Villegas [80], and Beukers, Cohen, Mellit [16] with a focus on motives defined over $\mathbb{Q}$. It offers a profound approach to hypergeometric motives and their L-functions by combining both classical and p-adic analysis, arithmetic, combinatorics, computation, and geometry. Our approach to the Galois perspective is derived from our attempt to understand hypergeometric functions over finite fields parallel to the classical setting and Weil's approach to consider Jacobi sums as Grössencharacters (or Grössencharakteres) [101]. This is reflected in our notation in Chapters 5 and 6 on the Galois perspective.

1.2. Organization and the main results

This memoir is organized as follows. Chapter 2 contains preliminaries on gamma and beta functions and their finite field counterparts, Gauss and Jacobi sums. In §2.4 we give a correspondence between objects in the classical and finite field settings. We recall some well-known facts for classical hypergeometric functions in Chapter 3. We assume in this memoir that the parameters for all the hypergeometric functions are rational numbers. In Chapter 4, we introduce the period functions $_{n+1}\mathbb{P}_n$ in Definition (4.4), and the normalized period functions $_{n+1}\mathbb{F}_n$ in Definition (4.9).

In Chapter 5, we provide some related background on Galois representations. In Chapter 6, we interpret the $_{n+1}\mathbb{P}_n$ and $_{n+1}\mathbb{F}_n$ functions, in particular the $n = 1$ case, as traces of Galois representations at Frobenius elements via explicit hypergeometric algebraic varieties. Let $\zeta_N := e^{2\pi i/N}$ be a primitive Nth root of 1 and $K = \mathbb{Q}(\zeta_N)$. We use $\mathcal{O}_K$ for its ring of integers, $\overline{K}$ for its algebraic closure, and set $G_K := \text{Gal}(\overline{K}/K)$. Recall that a prime ideal $\mathfrak{p}$ of $\mathcal{O}_K$ is unramified if it is coprime to the discriminant of K. Given a fixed rational number of the form $\frac{i}{N}$, for any nonzero prime ideal $\mathfrak{p}$ of K coprime to N, one can assign a multiplicative character, denoted by $\iota_\mathfrak{p}(\frac{i}{N})$, on the residue field $\mathcal{O}_K/\mathfrak{p}$ of $\mathfrak{p}$ with size $q(\mathfrak{p}) := \#(\mathcal{O}_K/\mathfrak{p})$, see (5.3). This assignment, based on the Nth power residue symbol, is *compatible* with the Galois perspective when $\mathfrak{p}$ varies, by which we mean one can build continuous Galois representations using the residue symbol. Then our setting for the finite field period functions and the dictionary between the two settings give us one way to convert the classical hypergeometric functions to the hypergeometric functions over the residue fields in a compatible way in the same sense mentioned above. We

1.2. ORGANIZATION AND THE MAIN RESULTS

illustrate the map by the following diagram

$$_{n+1}P_n \begin{bmatrix} a_1 & a_2 & \cdots & a_{n+1} \\ & b_1 & \cdots & b_n \end{bmatrix}; \lambda$$

$$\mapsto\ _{n+1}\mathbb{P}_n \begin{bmatrix} \iota_{\mathfrak{p}}(a_1) & \iota_{\mathfrak{p}}(a_2) & \cdots & \iota_{\mathfrak{p}}(a_{n+1}) \\ & \iota_{\mathfrak{p}}(b_1) & \cdots & \iota_{\mathfrak{p}}(b_n) \end{bmatrix}; \lambda; q(\mathfrak{p}) \Bigg],$$

where N is the least positive common denominator of all a_i's and b_j's, and $\mathfrak{p}$ is any prime ideal of $\mathcal{O}_K$ unramified in $K/\mathbb{Q}$. There is a similar correspondence between the hypergeometric functions $_{n+1}F_n$ and $_{n+1}\mathbb{F}_n$.

Thus when we speak of the finite field analogues of classical period (or hypergeometric) functions in this paper we mean the converted functions over the finite residue fields, unless we specify otherwise. We show the following.

THEOREM 1.1. *Let $a, b, c \in \mathbb{Q}$ with least common denominator N such that a, b, $a-c$, $b-c \notin \mathbb{Z}$ and $\lambda \in \mathbb{Q} \setminus \{0, 1\}$. Set $K = \mathbb{Q}(\zeta_N)$ and denote its ring of integers $\mathcal{O}_K$. Let ℓ be any prime. Then there exists a representation*

$$\sigma_{\lambda,\ell} : G_K := Gal(\overline{K}/K) \to GL_2(\mathbb{Q}_\ell(\zeta_N)),$$

depending on a, b and c, that is unramified at all prime ideals $\mathfrak{p}$ of $\mathcal{O}_K$ that are relatively prime to $\ell \cdot Disc(K/\mathbb{Q})$ and satisfy $ord_{\mathfrak{p}}(\lambda) = 0 = ord_{\mathfrak{p}}(1 - \lambda)$. Furthermore, the trace of Frobenius at $\mathfrak{p}$ in the image of $\sigma_{\lambda,\ell}$ is the well-defined algebraic integer

$$(1.2) \qquad -_2\mathbb{P}_1 \begin{bmatrix} \iota_{\mathfrak{p}}(a) & \iota_{\mathfrak{p}}(b) \\ & \iota_{\mathfrak{p}}(c) \end{bmatrix}; \lambda; q(\mathfrak{p}) \Bigg].$$

Our proof (see §6.3) makes use of generalized Legendre curves (recalled in §6.2) which are 1-dimensional hypergeometric varieties. Viewing the finite field period functions as the traces of finite dimensional Galois representations at Frobenius elements is helpful when we translate classical results to the finite field setting. Using Weil's result ([**101**]) realizing Jacobi sums as Grössencharacters, one can give a similar interpretation for the normalized period functions $_{n+1}\mathbb{F}_n$. In Chapter 7 we use the Galois perspective to discuss a finite field analogue of the Clausen formula by Evans and Greene, as well as its relations to analogues of some Ramanujan type formulas for $1/\pi$. For readers only interested in direct finite field analogues of classical formulas, Chapters 5, 6 and 7 can be skipped as our later proofs (except in Chapter 10) are mainly about character sums over finite fields.

In Chapter 8 we first recall and reinterpret the finite field analogues of Kummer's 24 relations considered by Greene [**41**]. These relations give us many built-in symmetries for the $_2F_1$ functions, which are closely related to the classical Kummer relations for $_2F_1$ (see §3.2.3). Then we use the dictionary between the classical and finite field settings to obtain a few finite field analogues of algebraic hypergeometric identities. For example, we recall the following formula of Slater (see [**88**, (1.5.20)]):

$$(1.3) \qquad _2F_1 \begin{bmatrix} a & a - \frac{1}{2} \\ & 2a \end{bmatrix}; z \Bigg] = \left(\frac{1 + \sqrt{1-z}}{2} \right)^{1-2a}.$$

Here, we call such a formula, which expresses a hypergeometric function (formally) as an algebraic function of the argument z, an *algebraic hypergeometric identity*. When $a \in \mathbb{Q}$, the corresponding monodromy group (§3.2.4) is a finite group. The identity (1.3) also satisfies the (∗) condition of §1.1. To see its finite field analogue, it is tempting to translate the right hand side into a corresponding character evaluated at $\frac{1+\sqrt{1-z}}{2}$ using the correspondence in §2.4. However, Theorem 1.1 implies that this type of direct translation will not yield a correct formula, as the corresponding Galois representations should be 2-dimensional. In fact, we will show (see Theorem 8.11) that for $\mathbb{F}_q$ of odd characteristic, ϕ the quadratic character, A any multiplicative character on $\mathbb{F}_q^\times$ having order at least 3, and $z \in \mathbb{F}_q$,

$$ {}_2\mathbb{F}_1\left[\begin{matrix} A & A\phi \\ & A^2 \end{matrix}; z\right] = \left(\frac{1+\phi(1-z)}{2}\right)\left(\overline{A}^2\left(\frac{1+\sqrt{1-z}}{2}\right) + \overline{A}^2\left(\frac{1-\sqrt{1-z}}{2}\right)\right). $$

Note the right hand side is well-defined as its first factor takes value zero if $1-z$ is not a square in $\mathbb{F}_q$ and 1 otherwise. Here we are using a bar to denote complex conjugation. This result was inspired by a recent result of Tu and Yang in [**93**]. It implies the following formula. Writing $\widehat{\mathbb{F}_q^\times}$ for the set of all multiplicative characters on $\mathbb{F}_q^\times$, if $A, B, AB, A\overline{B} \in \widehat{\mathbb{F}_q^\times}$ each have order larger than 2, then for $z \ne 1$,

$$ {}_2\mathbb{F}_1\left[\begin{matrix} A & A\phi \\ & A^2 \end{matrix}; z\right] {}_2\mathbb{F}_1\left[\begin{matrix} B & B\phi \\ & B^2 \end{matrix}; z\right] $$
$$ = {}_2\mathbb{F}_1\left[\begin{matrix} AB & AB\phi \\ & (AB)^2 \end{matrix}; z\right] + \overline{B}^2\left(\frac{z}{4}\right) {}_2\mathbb{F}_1\left[\begin{matrix} A\overline{B} & A\overline{B}\phi \\ & (A\overline{B})^2 \end{matrix}; z\right]. $$

Note that the last term is necessarily symmetric in A and B, although this is not obvious from its appearance.

In Chapter 9, we apply our main technique, which is to obtain analogues of classical formulas satisfying the (∗) condition using the dictionary between the two settings, to prove a quadratic formula over finite fields (Theorem 9.4). This is equivalent to a quadratic formula of Greene in [**41**]. The proof appears technical on the surface, but is parallel to the classical proof. In comparison, the approaches of Evans and Greene to higher order transformation formulas (such as [**28**, **41**]) often involve subtle and clever changes of variables.

Using a similar approach, we prove analogues of the Bailey cubic ${}_3F_2$ formulas (see Theorems 9.13 and 9.20) and consequently an analogue of a formula by Andrews and Stanton, see Theorem 9.12.

Next, we use a different approach to obtain a finite field analogue of the following cubic formula ([**38**, (5.18)]) by Gessel and Stanton. For $a, x \in \mathbb{C}$,

$$ {}_2F_1\left[\begin{matrix} a & -a \\ & \frac{1}{2} \end{matrix}; \frac{27x(1-x)^2}{4}\right] = {}_2F_1\left[\begin{matrix} 3a & -3a \\ & \frac{1}{2} \end{matrix}; \frac{3x}{4}\right], $$

when both sides converge. This formula satisfies the (∗) condition. We obtain the following explicit finite field analogue.

THEOREM 1.2. *Let $\mathbb{F}_q$ be the finite field of q elements and assume its residue characteristic is at least 3. If $A \in \widehat{\mathbb{F}_q^\times}$, then for all $x \in \mathbb{F}_q$,*

$$ {}_2\mathbb{F}_1\left[\begin{matrix} A & \overline{A} \\ & \phi \end{matrix}; \frac{27x(1-x)^2}{4}\right] = $$
$$ {}_2\mathbb{F}_1\left[\begin{matrix} A^3 & \overline{A}^3 \\ & \phi \end{matrix}; \frac{3x}{4}\right] - \phi(-3)\delta(x-1) - \phi(-3)A(-1)\delta(x-4/3), $$

where ϕ is the quadratic character, $\delta(0) = 1$ and $\delta(x) = 0$ otherwise.

As a consequence of this cubic formula, we obtain the following *evaluation formula*, by which we mean it is a formula which gives the value of a hypergeometric series at a fixed argument. In the theorem statement below $J(\,,\,)$ and $B(\,,\,)$ denote the Jacobi sum and beta function. We will recall their definitions in §2.1 and §2.2. As we shall see later, the Jacobi sums are the finite field analogues of the beta functions.

THEOREM 1.3. *Assume that q is a prime power with $q \equiv 1 \pmod{6}$, and let $A, \chi, \eta_3 \in \widehat{\mathbb{F}_q^\times}$ be such that η_3 has order 3, and none of $A^6, \chi^6, (A\chi)^3, (\overline{A}\chi)^3$ is the trivial character. Then,*
$$_3\mathbb{F}_2 \begin{bmatrix} A^3 & \overline{A}^3 & \chi \\ & \phi & \overline{\chi}^3 \end{bmatrix}; \frac{3}{4} = \sum_{\substack{B \in \widehat{\mathbb{F}_q^\times} \\ B^3 = A^3}} \frac{J(B\chi, \eta_3)J(\overline{B}\chi, \overline{\eta_3})}{J(B, \chi\eta_3)J(\overline{B}, \chi\overline{\eta_3})} = A(-1) \sum_{B^3 = A^3} \frac{J(B\chi, \overline{B}\chi)}{J(\chi\eta_3, \chi\overline{\eta_3})}.$$

This is an analogue of the following result of Gessel and Stanton [**38**] for $n \in \mathbb{N}$, $a \in \mathbb{C}$,

(1.4) $\quad _3F_2 \begin{bmatrix} 1+3a & 1-3a & -n \\ & \frac{3}{2} & -1-3n \end{bmatrix}; \frac{3}{4} = \frac{B(1+a+n, \frac{2}{3})B(1-a+n, \frac{4}{3})}{B(1+a, n+\frac{2}{3})B(1-a, n+\frac{4}{3})}.$

It is worth mentioning that Gessel and Stanton's proof of (1.4) uses a derivative, and thus cannot be translated directly to the finite field setting using our first technique. We include this case here to make the following point. The Galois perspective can provide helpful guidance in general, beyond using the $(*)$ condition.

In Chapter 10, we give an explicit application of finite field formulas in point counting on hypergeometric varieties. To be more explicit, we apply Theorem 9.4, a finite field quadratic transformation, to obtain the decomposition of a generically 4-dimensional abelian variety arising naturally from the generalized Legendre curve $y^{12} = x^9(1-x)^5(1-\lambda x)$, (see Theorem 10.1).

In Chapter 11, we conclude with a summary and some open questions based on numerical evidence.

In summary, our setup and main techniques allow us to make translations from the classical setting to the finite field setting in a straightforward manner, and the Galois perspective gives us additional helpful guidelines. Furthermore, these formulas over finite fields can be used to better understand and further compute global arithmetic invariants of the corresponding hypergeometric abelian varieties. We have to be careful: if a result requires additional structures in its proof such as derivative structure or the Lagrange inversion formula, for example in the case of the Pascal triangle identity $\binom{n}{m} + \binom{n}{m-1} = \binom{n+1}{m+1}$, then we cannot expect a direct translation.[1] Rather, our approaches are more suitable for translating results with geometric/motivic interpretation. In this sense, we will give a few finite field analogues of Ramanujan type formulas for $1/\pi$, see §7.2.

REMARK 1.4. We note that our definition of the finite field hypergeometric function is inspired by the advantages afforded by both the definitions of Greene

---

[1] In [**31**] by Evans, there is an analogue for the derivative operator over finite fields ([**31**, (2.8)]) which satisfies an analogue of the Leibniz rule ([**31**, Theorem 2.2]). However it is not a derivation. In [**42**], Greene gives the Lagrange inversion formula over finite fields, see Theorem 2.10 below, and points out its drawbacks.

[**41**] and McCarthy [**70**]. We make a small adjustment with the purpose of keeping the benefits of both of these while also closely aligning with the classical context and remaining compatible with the Galois representation perspective. We note, in particular, that our choice of notation implies that results can be stated for all values of x in $\mathbb{F}_q$ and we can interchange parameters to streamline alignment with classical proofs as much as possible. For a comparison of our definition with the others, see §4.4.

This monograph is an extension of our previous hybrid approach using both finite field and Galois methods. In [**23**], Deines et al. study Jacobian varieties arising from generalized Legendre curves. In [**24**], the authors use finite field hypergeometric evaluation formulas to compute the local zeta functions of some higher dimensional hypergeometric varieties. In [**36**] the authors consider the Appell-Lauricella hypergeometric functions over finite fields while in the more recent paper [**59**] the authors obtain a Whipple $_7F_6(1)$ formula over finite fields and use it to construct decomposable Galois representations. This information allows us to explain the geometry behind several supercongruences satisfied by the corresponding truncated hypergeometric functions. These supercongruences were proved using a method in [**68**] by Long and Ramakrishna which is based on the local analytic properties of the p-adic Gamma functions. The approaches to supercongruences involve tools such as hypergeometric functions over finite fields as in [**3**] by Ahlgren and Ono and [**76**] by Osburn and Schneider, [**69**] by Long, Tu, Yui and Zudilin, [**67**] by Long, or the Wiff-Zeilberger method as in [**108**] by Zudilin and [**77**] by Osburn and Zudilin.

CHAPTER 2

Preliminaries for the Complex and Finite Field Settings

In this chapter, we describe the foundation for the finite field analogue of the hypergeometric function. To do so, we first recall preliminaries from the classical setting. For more details on the classical setting see [6, 11, 88], and see [14, 31, 32, 41] for the finite field setup.

2.1. Gamma and beta functions

Most of the details in this section can be found in Chapter 1 of Andrews, Askey and Roy, [6]. We let $\mathbb{N} = \mathbb{Z}_{>0}$ throughout.

Recall the usual binomial coefficient
$$\binom{n}{k} = \frac{n!}{k!(n-k)!}.$$

For any $a \in \mathbb{C}$ and $n \in \mathbb{N}$, define the rising factorial by

(2.1) $$(a)_n := a(a+1)\cdots(a+n-1),$$

and define $(a)_0 = 1$. Note then, that

(2.2) $$\frac{(-1)^k(-n)_k}{k!} = \binom{n}{k}.$$

The *gamma function* $\Gamma(x)$ and *beta function* $B(x,y)$ are defined as follows.

DEFINITION 2.1. For $\operatorname{Re}(x) > 0$,
$$\Gamma(x) := \int_0^\infty t^{x-1} e^{-t} \, dt.$$

While $\Gamma(x)$ can be extended to a meromorphic function with poles at non-positive integers, its reciprocal $1/\Gamma(x)$ is an entire function and, like $\sin(x)$, has a product representation (see Theorem 1.1.2 of [6]). The gamma function satisfies the functional equation

(2.3) $$\Gamma(x+1) = x\Gamma(x),$$

which can easily be derived using integration by parts. By (2.3), one has that for $n \in \mathbb{Z}_{\geq 0}$,

(2.4) $$(a)_n = \frac{\Gamma(a+n)}{\Gamma(a)}.$$

DEFINITION 2.2. For $\operatorname{Re}(x) > 0, \operatorname{Re}(y) > 0$

$$B(x,y) := \int_0^1 t^{x-1}(1-t)^{y-1}\,dt.$$

The assumptions on x and y can be relaxed by integrating along the Pochhammer contour path around 0 and 1 defined as follows.

DEFINITION 2.3. Let a, b be two points in $\mathbb{C}P^1$. Each Pochhammer contour γ_{ab} is a closed curve corresponding to a commutator of the form $ABA^{-1}B^{-1}$ in the fundamental group of $\pi_1(\mathbb{C}P^1 \setminus \{a,b,\infty\})$, where $A, B \in \pi_1(\mathbb{C}P^1 \setminus \{a,b,\infty\})$ are loops around both of the points a, b and the superscript -1 denotes a path taken in the opposite direction.

For example, when $a = 0$, $b = 1$, a contour γ_{01} is a closed curve starting from a fixed point $T \in (0,1)$, going around 0 and 1 counterclockwise (in that order) and returning to T. Then one loops around 0 and 1 clockwise returning again to T, as indicated by the picture below.

Integrating over the double contour loop γ_{01}, the integral

$$B(x,y) = \frac{1}{(1-e^{2\pi i x})(1-e^{2\pi i y})} \int_{\gamma_{01}} t^{x-1}(1-t)^{y-1} dt$$

converges for all values of x and y. For details, see [48, 106, 107]. The beta function satisfies the functional equation

(2.5) $$B(x,y) = \frac{x+y}{y} B(x, y+1),$$

which can also be obtained using integration by parts (see (1.1.14) in [6]).

Using (2.5), one can prove (see Theorem 1.1.4 in [6]) that the gamma and beta functions are related via

(2.6) $$B(x,y) = \frac{\Gamma(x)\Gamma(y)}{\Gamma(x+y)}.$$

For $a \in \mathbb{C}$ and a variable z, the binomial theorem can be restated in the notation of the rising factorial as

(2.7) $$(1-z)^a = \sum_{k=0}^{\infty} \binom{a}{k}(-z)^k = \sum_{k=0}^{\infty} \frac{(-a)_k}{k!} z^k.$$

Binomial coefficients, rising factorials, gamma and beta functions are all used frequently in the classical theory of hypergeometric functions. In fact, gamma and beta functions play more primary roles while binomial coefficients and rising factorials make the notation more compact. Also of fundamental importance are specific formulas involving the gamma function, which we now state. First, we give Euler's reflection formula.

THEOREM 2.4 (Euler's Reflection Formula, Theorem 1.2.1 of [6]). *For $a \in \mathbb{C}$,*
$$\Gamma(a)\Gamma(1-a) = \frac{\pi}{\sin(\pi a)}.$$

We note that Theorem 2.4 implies $\Gamma(\frac{1}{2}) = \sqrt{\pi}$ and, for example, that $\Gamma(\frac{1}{3})\Gamma(\frac{2}{3}) = \frac{2\sqrt{3}\pi}{3}$.

The gamma function also satisfies multiplication formulas.

THEOREM 2.5 (Legendre's Duplication Formula, Theorem 1.5.1 of [6]). *For $a \in \mathbb{C}$,*
$$\Gamma(2a)(2\pi)^{1/2} = 2^{2a-\frac{1}{2}}\Gamma(a)\Gamma\left(a + \frac{1}{2}\right).$$

Stated in terms of rising factorials, Theorem 2.5 gives that for all $n \in \mathbb{N}$,

(2.8) $$(a)_{2n} = 2^{2n} \left(\frac{a}{2}\right)_n \left(\frac{a+1}{2}\right)_n.$$

One way to prove Theorem 2.5 is the following. Using the functional equation (2.3) for $\Gamma(x)$, one can see that
$$h(x) := 2^{2x-1} \frac{\Gamma(x)\Gamma(x+\frac{1}{2})}{\Gamma(\frac{1}{2})\Gamma(2x)}$$
satisfies the functional equation
$$h(x+1) = h(x).$$
The result can then be derived using Stirling's formula, which describes the asymptotic behavior of $\Gamma(x)$ (see Remark 1.5.1 of [6]).

The more general case is Gauss' multiplication formula, given below.

THEOREM 2.6 (Gauss' Multiplication Formula, Theorem 1.5.2 of [6]). *For $m \in \mathbb{N}$ and $a \in \mathbb{C}$,*
$$\Gamma(ma)(2\pi)^{(m-1)/2} = m^{ma-\frac{1}{2}}\Gamma(a)\Gamma\left(a+\frac{1}{m}\right)\cdots\Gamma\left(a+\frac{m-1}{m}\right).$$

Stated in terms of rising factorials, for any $n \in \mathbb{N}$,
$$(ma)_{mn} = m^{mn} \prod_{i=0}^{m-1} \left(a + \frac{i}{m}\right)_n.$$

We end this subsection with the Lagrange inversion theorem for formal power series. If $f(z)$ and $g(z)$ are formal power series where $g(0) = 0$ and $g'(0) \neq 0$, then Lagrange's inversion theorem gives a way to write f as a power series in $g(z)$. In particular, one can write

(2.9) $$f(z) = f(0) + \sum_{k=1}^{\infty} c_k g(z)^k,$$

where
$$c_k = \operatorname{Res}_z \frac{f'(z)}{kg(z)^k},$$
and $\operatorname{Res}_z(f)$ denotes the coefficient of $1/z$ in the power series expansion of f. Lagrange's original result can be found in [54], and [39] by Good provides a simple

proof, as well as generalizations to several variables. We discuss finite field analogues of this result in §2.3.

2.2. Gauss and Jacobi sums

We now lay the necessary corresponding groundwork for the finite field setting based on [6, 14, 29, 30, 32, 41]. We recall the Gauss and Jacobi sums, which are the finite field analogues of the gamma and beta functions, and consequently the binomial coefficient and the rising factorial. We also state analogues of the formulas for $\Gamma(x)$ given in Theorems 2.4-2.6. For a comprehensive treatment of Gauss sums and Jacobi sums, see the book by Berndt, Evans and Williams [14].

To begin, fix an odd prime p and let $\mathbb{F}_q$ be a finite field of size q, where $q = p^e$. Recall that a *multiplicative character* χ on $\mathbb{F}_q^\times$ is a group homomorphism

$$\chi : \mathbb{F}_q^\times \to \mathbb{C}^\times,$$

and the set $\widehat{\mathbb{F}_q^\times}$ of all multiplicative characters on $\mathbb{F}_q^\times$ forms a cyclic group of order $q - 1$ under multiplication. Throughout, we fix the following notation

$$\varepsilon = \text{ the trivial character in } \widehat{\mathbb{F}_q^\times}$$

$$\phi = \text{ the quadratic character in } \widehat{\mathbb{F}_q^\times},$$

so that $\varepsilon(a) = 1$ for all $a \neq 0$, and ϕ is nontrivial such that $\phi^2 = \varepsilon$. We extend the definition of each character $\chi \in \widehat{\mathbb{F}_q^\times}$ to all of $\mathbb{F}_q$ by setting $\chi(0) = 0$, including $\varepsilon(0) = 0$.[1]

Further, following Greene in [41], it is useful to define a function δ for $\chi \in \widehat{\mathbb{F}_q^\times}$, or $x \in \mathbb{F}_q$, respectively, by

$$\delta(\chi) := \delta_\varepsilon(\chi) := \begin{cases} 1 & \text{if } \chi = \varepsilon, \\ 0 & \text{if } \chi \neq \varepsilon; \end{cases}$$

$$\delta(x) := \delta_0(x) := \begin{cases} 1 & \text{if } x = 0, \\ 0 & \text{if } x \neq 0. \end{cases}$$

This definition will allow us to describe formulas which hold for all characters, without having to separate cases involving trivial characters.

Let $\zeta_N := e^{2\pi i/N}$. We fix a primitive p^{th} root of unity ζ_p and $A, B \in \widehat{\mathbb{F}_q^\times}$. We define the *Gauss sum* and *Jacobi sum*, respectively as follows.

$$g(A) := \sum_{x \in \mathbb{F}_q^\times} A(x) \zeta_p^{\operatorname{Tr}_{\mathbb{F}_p}^{\mathbb{F}_q}(x)},$$

$$J(A, B) := \sum_{x \in \mathbb{F}_q} A(x) B(1 - x),$$

where $\operatorname{Tr}_{\mathbb{F}_p}^{\mathbb{F}_q}(x) := x + x^p + x^{p^2} + \cdots + x^{p^{e-1}}$ is the trace of x viewed as a surjective linear map from $\mathbb{F}_q$ to $\mathbb{F}_p$. It is not hard to see that

$$g(\varepsilon) = -1,$$

---

[1] A different choice would be $\varepsilon(0) = 1$, see page 9 of [14]. However, we would prefer to work with functions that are multiplicative for all elements in $\mathbb{F}_q$. Thus $\varepsilon(0) = 0$ is preferred.

a fact we will make use of frequently. While in general the Gauss sums depend on the choice of primitive root ζ_p, the Jacobi sums do not. By definition, $J(A, B)$ is in the ring of integers of the cyclotomic field $\mathbb{Q}(\zeta_{q-1})$, while $g(A)$ is in $\mathbb{Q}(\zeta_{p(q-1)})$.

For a more detailed introduction to characters, see [14] or Chapter 8 of [46]. We note that although [14] and [46] contain proofs of most results we give throughout the rest of this subsection, care must be taken for cases involving the trivial character ε, which is defined to be 1 at 0 in both references.

The Gauss sum is the finite field analogue of the gamma function, while the Jacobi sum is the finite field analogue of the beta function [32]. With this in mind, we observe that the following are finite field analogues of Euler's reflection formula from Theorem 2.4.

(2.10)
$$g(A)g(\overline{A}) = qA(-1) - (q-1)\delta(A)$$
$$\frac{1}{g(A)g(\overline{A})} = \frac{A(-1)}{q} + \frac{q-1}{q}\delta(A).$$

Note that the second equation of (2.10) can be derived from the first by multiplying both right hand sides together to obtain 1. Furthermore, since $g(\varepsilon) = -1$, we can write

(2.11)
$$g(\overline{A}) = \frac{qA(-1)}{g(A)} + (q-1)\delta(A)$$
$$\frac{1}{g(\overline{A})} = \frac{A(-1)g(A)}{q} - \frac{q-1}{q}\delta(A).$$

The finite field analogues of the multiplication formulas for $\Gamma(x)$ given in Theorems 2.5 and 2.6 are known as the Hasse-Davenport Relations. The general version given below is a finite field analogue of Theorem 2.6.

THEOREM 2.7 (Hasse-Davenport Relation, see Theorem 11.3.5 of [14]). *Let $m \in \mathbb{N}$ and $q = p^e$ be a prime power with $q \equiv 1 \pmod{m}$. For any multiplicative character $\psi \in \widehat{\mathbb{F}_q^\times}$, we have*

$$\prod_{\substack{\chi \in \widehat{\mathbb{F}_q^\times} \\ \chi^m = \varepsilon}} g(\chi\psi) = -g(\psi^m)\psi(m^{-m}) \prod_{\substack{\chi \in \widehat{\mathbb{F}_q^\times} \\ \chi^m = \varepsilon}} g(\chi).$$

The proof in [14] for the Hasse-Davenport relation is based on Stickelberger's congruence for Gauss sums stated in Section 11.2 of [14] or Section 3.7 of [22].[2]

We can obtain the analogue of Legendre's duplication formula in Theorem 2.5 by taking the case $m = 2$ of Theorem 2.7. This gives that for any multiplicative character A

(2.12)
$$g(A)g(\phi A) = g(A^2)g(\phi)\overline{A}(4),$$

where ϕ is the quadratic character as before.

Just as there is the relationship between the gamma and beta functions, (2.6), one can relate Gauss sums and Jacobi sums by

---

[2]H. Cohen pointed out to us that by equation (2.10), the product $-\prod_{\chi \in \widehat{\mathbb{F}_q^\times}, \chi^m = \varepsilon} g(\chi)$ is necessarily a quadratic character when m is odd; otherwise it is a quadratic character times $g(\phi)$, the quadratic Gauss sum.

$$(2.13) \qquad J(A,B) = \frac{g(A)g(B)}{g(AB)} + (q-1)B(-1)\delta(AB).$$

The left hand side of equation (2.13) is symmetric in A and B while there is an apparent asymmetry on the right hand side. Note that $\delta(AB) \neq 0$ means $B = \overline{A}$ and thus $B(-1) = \overline{A}(-1) = A(-1)$ so the right hand side is in fact symmetric.

Below, we note some special cases for Jacobi sums. We have,

$$(2.14) \qquad J(A,\overline{A}) = -A(-1) + (q-1)\delta(A)$$

and

$$(2.15) \qquad J(A,\varepsilon) = -1 + (q-1)\delta(A).$$

In [**14**] (see Theorem 2.1.4) there is an elementary proof of (2.12) written in terms of Jacobi sums. It gives, for $A \neq \varepsilon$,

$$(2.16) \qquad J(A,A) = \overline{A}(4)J(A,\phi).$$

We now are ready to define finite field analogues for the rising factorial and binomial coefficient[3] in (2.1) and (2.2). We let

$$(2.17) \qquad (A)_\chi := \frac{g(A\chi)}{g(A)},$$

$$(2.18) \qquad \binom{A}{\chi} := -\chi(-1)J(A,\overline{\chi}).$$

Then we see directly that for $A, \chi_1, \chi_2 \in \widehat{\mathbb{F}_q^\times}$,

$$(2.19) \qquad (A)_{\chi_1\chi_2} = (A)_{\chi_1}(A\chi_1)_{\chi_2},$$

which is the analogue of the classical identity $(a)_{n+m} = (a)_n(a+n)_m$, for integers $n, m \geq 0$. The multiplication formula in Theorem 2.7 can be written in this notation as

$$(2.20) \qquad (A^m)_{\psi^m} = \psi(m^m) \prod_{\substack{\chi \in \widehat{\mathbb{F}_q^\times} \\ \chi^m = \varepsilon}} (A\chi)_\psi,$$

which compares with the second statement in Theorem 2.6.

Now we give a proof of a useful fact about Jacobi sums (see (2.6) of Greene [**41**]), which is the analogue of the classical binomial identity

$$\binom{n}{m} = \binom{n}{n-m}.$$

LEMMA 2.8. *For any characters A, B on $\mathbb{F}_q^\times$, $J(A,\overline{B}) = A(-1)J(A,B\overline{A})$. Stated using binomial notation,*

$$\binom{A}{B} = \binom{A}{A\overline{B}}.$$

---

[3] Note that our binomial coefficient differs from Greene's by a factor of $-q$.

PROOF. We observe

$$J(A, \overline{B}) \stackrel{(2.13)}{=} \frac{g(A)g(\overline{B})}{g(A\overline{B})} + (q-1)B(-1)\delta(A\overline{B})$$

$$\stackrel{(2.11)}{=} \frac{g(A)}{g(A\overline{B})}\left(\frac{qB(-1)}{g(B)} + (q-1)\delta(B)\right) + (q-1)B(-1)\delta(A\overline{B})$$

$$\stackrel{(2.11)}{=} \frac{g(A)qB(-1)}{g(B)}\left(\frac{AB(-1)}{q}g(\overline{A}B) - \frac{q-1}{q}\delta(\overline{A}B)\right)$$
$$+ (q-1)\delta(B) + (q-1)B(-1)\delta(A\overline{B})$$

$$= A(-1)\frac{g(A)g(\overline{A}B)}{g(B)} + (q-1)\delta(B)$$

$$\stackrel{(2.13)}{=} A(-1)J(A, \overline{A}B).$$

$\square$

REMARK 2.9. The proof of Lemma 2.8 demonstrates that one strategy for proving identities is to use the basic properties of Gauss and Jacobi sums while keeping track of the delta terms. Alternatively, one can reprove the above by changing variables

$$A(-1)J(A, \overline{A}B) = A(-1)\sum_{x \in \mathbb{F}_q^\times} A(1-x)\overline{A}B(x)$$

$$= A(-1)\sum_{x \in \mathbb{F}_q^\times} A\left(\frac{1-x}{x}\right)B(x) = A(-1)\sum_{x \in \mathbb{F}_q^\times} A\left(\frac{1-x}{x}\right)\overline{B}\left(\frac{1}{x}\right)$$

$$= \sum_{x \in \mathbb{F}_q^\times} A\left(1 - \frac{1}{x}\right)\overline{B}\left(\frac{1}{x}\right) = J(A, \overline{B}).$$

The technique of changing variables makes the proofs more compact when used properly.

2.3. Lagrange inversion

Below is a finite field analogue of the Lagrange inversion formula (2.9), which is similar to the Fourier inversion formula given in [**16**]. We state the version where the basis of complex valued functions on the finite field is comprised of all multiplicative characters in $\widehat{\mathbb{F}_q^\times}$, together with $\delta(x)$.

THEOREM 2.10 ([**42**] Theorem 2.7). *Let p be an odd prime, $q = p^e$, and suppose $f : \mathbb{F}_q \to \mathbb{C}$ and $g : \mathbb{F}_q \to \mathbb{F}_q$ are functions. Then*

$$\sum_{\substack{y \in \mathbb{F}_q \\ g(y) = g(x)}} f(y) = \delta(g(x))\sum_{\substack{y \in \mathbb{F}_q \\ g(y) = 0}} f(y) + \sum_{\chi \in \widehat{\mathbb{F}_q^\times}} f_\chi \chi(g(x)),$$

where

$$f_\chi = \frac{1}{q-1}\sum_{y \in \mathbb{F}_q} f(y)\overline{\chi}(g(y)).$$

Compared with the classical formula (2.9), the assumptions $f(0) = 0$, $f'(0) \neq 0$, i.e. the map being one-to-one near 0, are not required. Greene pointed out that it is also the reason why the finite field version cannot be used to determine coefficients when f is not a one-to-one function.

We will use Theorem 2.10 in Chapter 4 to develop the finite field analogue of the hypergeometric function.

2.4. A dictionary between the complex and finite field settings

We now list a dictionary that we will use for convenience. It is well-known to experts. Let $N \in \mathbb{N}$, and $a, b \in \mathbb{Q}$ with common denominator N.

$$\begin{array}{ccc}
\frac{1}{N} & \leftrightarrow & \text{an order } N \text{ character } \eta_N \in \widehat{\mathbb{F}_q^\times} \\
a = \frac{i}{N},\ b = \frac{j}{N} & \leftrightarrow & A, B \in \widehat{\mathbb{F}_q^\times},\ A = \eta_N^i,\ B = \eta_N^j \\
x^a & \leftrightarrow & A(x) \\
x^{a+b} & \leftrightarrow & A(x)B(x) = AB(x) \\
a + b & \leftrightarrow & A \cdot B \\
-a & \leftrightarrow & \overline{A} \\
\Gamma(a) & \leftrightarrow & g(A) \\
(a)_n = \Gamma(a+n)/\Gamma(a) & \leftrightarrow & (A)_\chi = g(A\chi)/g(A) \\
B(a,b) & \leftrightarrow & J(A,B) \\
\int_0^1\ \mathrm{d}x & \leftrightarrow & \sum_{x \in \mathbb{F}} \\
\Gamma(a)\Gamma(1-a) = \frac{\pi}{\sin a\pi},\ a \notin \mathbb{Z} & \leftrightarrow & g(A)g(\overline{A}) = A(-1)q,\ A \neq \varepsilon \\
(ma)_{mn} = m^{mn} \prod_{i=1}^m \left(a + \frac{i}{m}\right)_n & \leftrightarrow & (A^m)_{\psi^m} = \psi(m^m) \prod_{i=1}^m (A\eta_m^i)_\psi
\end{array}$$

REMARK 2.11. One could use $\frac{1}{q}\sum_{x \in \mathbb{F}}$ for the finite field analogue of $\int_0^1 \mathrm{d}x$ so that the 'volume' $\frac{1}{q}\sum_{x \in \mathbb{F}} 1 = 1$. Another option is to use $\frac{1}{q-1}\sum_{x \in \mathbb{F}^\times}$. Our choice above is instead $\sum_{x \in \mathbb{F}}$, which makes point counting formulas more natural.

REMARK 2.12. It is useful to note here that since integers correspond to the trivial character, integers appearing in exponents will disappear when translating to the finite field setting.

CHAPTER 3

Classical Hypergeometric Functions

In this chapter, we recall the definition and basic properties of classical hypergeometric functions, including transformation and evaluation formulas. We also discuss the hypergeometric differential equation and its relationship to the Schwarz map and the monodromy group.

3.1. Classical development

We now recall the basic development of the classical (generalized) hypergeometric functions. See [6, 11, 88] for a thorough treatment. For all classical formulas stated in this paper, the assumption is always that they hold when both sides of the formula are well-defined and converge.

The classical (generalized) *hypergeometric functions* $_{n+1}F_n$ with complex parameters $a_1, \ldots, a_{n+1}, b_1, \ldots, b_n$, and argument z are defined by

$$(3.1) \quad _{n+1}F_n \begin{bmatrix} a_1 & a_2 & \cdots & a_{n+1} \\ & b_1 & \cdots & b_n \end{bmatrix} ; z \end{bmatrix} := \sum_{k=0}^{\infty} \frac{(a_1)_k \cdots (a_{n+1})_k}{(b_1)_k \cdots (b_n)_k} \frac{z^k}{k!},$$

and converge when $|z| < 1$. Each $_{n+1}F_n$ function satisfies an order $n+1$ ordinary Fuchsian differential equation in the variable z with three regular singularities at 0, 1, and ∞ [88, §2.1.2][105]. It also satisfies the following inductive integral relation of Euler [6, Equation (2.2.2)]. Namely, when $\operatorname{Re}(b_n) > \operatorname{Re}(a_{n+1}) > 0$,

$$(3.2) \quad _{n+1}F_n \begin{bmatrix} a_1 & a_2 & \cdots & a_{n+1} \\ & b_1 & \cdots & b_n \end{bmatrix} ; z \end{bmatrix} = B(a_{n+1}, b_n - a_{n+1})^{-1}$$

$$\cdot \int_0^1 t^{a_{n+1}-1}(1-t)^{b_n - a_{n+1} - 1} \cdot {_nF_{n-1}} \begin{bmatrix} a_1 & a_2 & \cdots & a_n \\ & b_1 & \cdots & b_{n-1} \end{bmatrix} ; zt \end{bmatrix} dt.$$

To relate these functions to periods of algebraic varieties (see [52] for useful discussions on periods), we first observe that by (2.7) we may define for general $a \in \mathbb{C}$,

$$(3.3) \quad _1P_0[a; z] := (1-z)^{-a} = \sum_{k=0}^{\infty} \frac{(a)_k}{k!} z^k = {_1F_0}[a; z].$$

Here we are using the notation $_1P_0$ to indicate a relationship to periods of algebraic varieties when $a \in \mathbb{Q}$. Next we let

$$(3.4) \quad _2P_1 \begin{bmatrix} a & b \\ & c \end{bmatrix} ; z \end{bmatrix} := \int_0^1 t^{b-1}(1-t)^{c-b-1} {_1P_0}[a; zt] \, dt$$

$$= \int_0^1 t^{b-1}(1-t)^{c-b-1}(1-zt)^{-a} dt,$$

or one can define it using the Pochhammer contour described in §2.1, see [**88**, §1.6]. Up to an algebraic multiple, the value of this function at a given z can be realized as a period of a corresponding generalized Legendre curve, see [**8, 23, 103**] and §6.2 below. To relate it to the $_2F_1$ function, one uses (3.2) to see that when $\operatorname{Re}(c) > \operatorname{Re}(b) > 0$,

$$(3.5) \quad {_2P_1}\begin{bmatrix} a & b \\ & c \end{bmatrix}; z\end{bmatrix} = \int_0^1 t^{b-1}(1-t)^{c-b-1} \cdot {_1P_0}[a; zt]dt$$

$$= \int_0^1 t^{b-1}(1-t)^{c-b-1} \cdot {_1F_0}[a; zt]dt$$

$$= B(b, c-b) \cdot {_2F_1}\begin{bmatrix} a & b \\ & c \end{bmatrix}; z\end{bmatrix}.$$

Inductively one can define the (higher) periods $_{n+1}P_n$ similarly by

$$(3.6) \quad {_{n+1}P_n}\begin{bmatrix} a_1 & a_2 & \cdots & a_{n+1} \\ & b_1 & \cdots & b_n \end{bmatrix}; z\end{bmatrix} :=$$

$$\int_0^1 t^{a_{n+1}-1}(1-t)^{b_n-a_{n+1}-1} \, {_nP_{n-1}}\begin{bmatrix} a_1 & a_2 & \cdots & a_n \\ & b_1 & \cdots & b_{n-1} \end{bmatrix}; zt\end{bmatrix} dt.$$

Again using the beta function, one can show that when $\operatorname{Re}(b_i) > \operatorname{Re}(a_{i+1}) > 0$ for each $i \geq 1$,

$$_{n+1}F_n\begin{bmatrix} a_1 & a_2 & \cdots & a_{n+1} \\ & b_1 & \cdots & b_n \end{bmatrix}; z\end{bmatrix} =$$

$$\prod_{i=1}^n B(a_{i+1}, b_i - a_{i+1})^{-1} \cdot {_{n+1}P_n}\begin{bmatrix} a_1 & a_2 & \cdots & a_{n+1} \\ & b_1 & \cdots & b_n \end{bmatrix}; z\end{bmatrix}.$$

By the definition of $_{n+1}F_n$ in (3.1), any given $_{n+1}F_n$ function satisfies two nice properties:

1) The leading coefficient is 1;
2) The roles of the upper entries a_i (resp. lower entries b_j) are symmetric.

Clearly, the $_{n+1}P_n$ period functions do not satisfy these properties in general. The hypergeometric functions can thus be viewed as periods that are 'normalized' so that both properties 1) and 2) are satisfied.

These two properties are important motivating factors for our definition of a finite field analogue of $_{n+1}F_n$ in §4.3.

3.2. Some properties of hypergeometric functions with $n = 1$

Here we focus on some important properties of the $_2F_1$ functions. For details see [**6, 15, 88**].

3.2.1. The hypergeometric differential equation.
First, we recall that the hypergeometric function

$$_2F_1\begin{bmatrix} a & b \\ & c \end{bmatrix}; z\end{bmatrix}$$

is a solution of the *hypergeometric differential equation*

$$(3.7) \quad HDE(a, b; c; z) : z(1-z)F'' + [c - (a+b+1)z]F' - abF = 0,$$

where here $'$ means derivative with respect to z. This is a Fuchsian differential equation with 3 regular singularities at 0, 1, and ∞ [**105**].

PROPOSITION 3.1. *The space of solutions to $HDE(a,b;c;z)$ has bases $\{\beta_z, \gamma_z\}$ given below, where β_z, γ_z are expanded near the singularity $z \in \{0, 1, \infty\}$.*

(1) *When $z = 0$, and $c \notin \mathbb{Z}$,*

$$\beta_0 := {}_2F_1 \begin{bmatrix} a & b \\ & c \end{bmatrix}; z$$

$$\gamma_0 := z^{1-c} {}_2F_1 \begin{bmatrix} 1+a-c & 1+b-c \\ & 2-c \end{bmatrix}; z.$$

(2) *When $z = 1$, and $a+b-c \notin \mathbb{Z}$,*

$$\beta_1 := {}_2F_1 \begin{bmatrix} a & b \\ & 1+a+b-c \end{bmatrix}; 1-z$$

$$\gamma_1 := (1-z)^{c-a-b} {}_2F_1 \begin{bmatrix} c-a & c-b \\ & 1+c-a-b \end{bmatrix}; 1-z.$$

(3) *When $z = \infty$, and $a-b \notin \mathbb{Z}$,*

$$\beta_\infty := z^{-a} {}_2F_1 \begin{bmatrix} a & 1+a-c \\ & 1+a-b \end{bmatrix}; 1/z$$

$$\gamma_\infty := z^{-b} {}_2F_1 \begin{bmatrix} b & 1+b-c \\ & 1+b-a \end{bmatrix}; 1/z.$$

We note that near the singularities $z = 0$, 1, or ∞, in the cases when $c \in \mathbb{Z}$, $a+b-c \in \mathbb{Z}$, or $a-b \in \mathbb{Z}$, respectively, the equation $HDE(a,b;c;z)$ has solutions with logarithmic terms near the corresponding singularity.

3.2.2. Schwarz map and Schwarz triangles. When $a, b, c \in \mathbb{Q}$, there is an explicit correspondence between a hypergeometric differential equation $HDE(a, b; c; z)$ and a Schwarz triangle $\Delta(p, q, r)$ with $p, q, r \in \mathbb{Q}$ due to the following theorem of Schwarz.

THEOREM 3.2 (Schwarz, (see [**105**])). *Let f, g be two independent solutions to the differential equation $HDE(a, b; c; z)$ at a point $z \in \mathfrak{H}$, the complex upper half-plane, and let $p = |1-c|$, $q = |c-a-b|$, and $r = |a-b|$. Then the Schwarz map $D = f/g$ gives a bijection from $\mathfrak{H} \cup \mathbb{R}$ onto a curvilinear triangle with vertices $D(0)$, $D(1)$, $D(\infty)$ and corresponding angles $p\pi$, $q\pi$, $r\pi$, as illustrated below. (Note that a change of the basis $\{f,g\}$ corresponds to a fractional linear transformation which does not change the angles of the curvilinear triangle).*

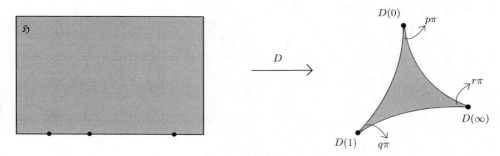

Note that a Schwarz triangle with angles $p\pi$, $q\pi$, and $r\pi$ as described in Theorem 3.2 can be used to tile the sphere, the Euclidean plane, or the hyperbolic plane through reflections along its edges, depending on whether $p+q+r$ is equal to, greater than, or less than 1, respectively. Therefore, each Schwarz triangle $\Delta(p,q,r)$ can be associated to the symmetry group of this tiling, which we denote by $S_\Delta(p,q,r)$.

3.2.3. Kummer relations. For generic z, $HDE(a,b;c;z)$ has group of symmetries $G(a,b;c;z)$ of order 24. This symmetry group acts projectively on its vector space of solutions. Let Σ be the set of all ordered triples (a,b,c). The group $G(a,b;c;z)$ is related both to the set of all 6 linear fractional transformations

$$\{z, 1-z, 1/z, z/(z-1), 1/(1-z), 1-1/z\}$$

that permutes $0, 1, \infty$, and to the set of finite order maps from Σ to itself preserving the (unordered) set of Schwarz angles $\{|1-c|, |c-a-b|, |a-b|\}$. For example, both $(a,b,c) \mapsto (c-a,b,c)$ and $(a,b,c) \mapsto (a,c-b,c)$ have order 2 and preserve the set of Schwarz angles. So does the composition map $(a,b,c) \mapsto (c-a,c-b,c)$.

The group $G(a,b;c;z)$ is isomorphic to the symmetric group S_4 (see [**53**], [**6**, §2.3]) and is an extension of S_3 (permuting the singular points $0, 1, \infty$), by the Klein 4-group V_4 via a short exact sequence

$$1 \to V_4 \to G(a,b;c;z) \to S_3 \to 1.$$

In particular, the action of V_4 is via the following two identities (see [**6**, Theorem 2.2.5]), which correspond to the order 2 maps on Σ above. The first one is due to Pfaff

$$(3.8) \quad {}_2F_1\begin{bmatrix} a & b \\ & c \end{bmatrix};z\bigg] = (1-z)^{-a}\,{}_2F_1\begin{bmatrix} a & c-b \\ & c \end{bmatrix};\frac{z}{z-1}\bigg]$$

$$= (1-z)^{-b}\,{}_2F_1\begin{bmatrix} c-a & b \\ & c \end{bmatrix};\frac{z}{z-1}\bigg],$$

and can be obtained from the integral representation of the ${}_2F_1$ and a change of variables. The second is Euler's formula

$$(3.9) \quad {}_2F_1\begin{bmatrix} a & b \\ & c \end{bmatrix};z\bigg] = (1-z)^{c-a-b}\,{}_2F_1\begin{bmatrix} c-a & c-b \\ & c \end{bmatrix};z\bigg],$$

which can be deduced from the Pfaff transformations in (3.8).

In addition to (3.8) and (3.9), the other Kummer relations describe the analytic continuation of hypergeometric functions at the other singularities. For example, the analytic continuation of

$$_2F_1\begin{bmatrix} a & b \\ & c \end{bmatrix}; z$$

from near 0 to ∞ can be expressed by following equation [6, Theorem 2.3.2],

$$(3.10) \quad _2F_1\begin{bmatrix} a & b \\ & c \end{bmatrix}; z = \frac{\Gamma(c)\Gamma(b-a)}{\Gamma(c-a)\Gamma(b)} \cdot (-z)^{-a} {}_2F_1\begin{bmatrix} a & a-c+1 \\ & a-b+1 \end{bmatrix}; 1/z$$
$$+ \frac{\Gamma(c)\Gamma(a-b)}{\Gamma(c-b)\Gamma(a)} \cdot (-z)^{-b} {}_2F_1\begin{bmatrix} b & b-c+1 \\ & b-a+1 \end{bmatrix}; 1/z.$$

For more details, see [6, 53, 88].

3.2.4. Monodromy. Consider $\mathbb{C}P^1 \setminus \{0,1,\infty\}$ as a topological space with a fixed base point t_0. The fundamental group $\pi_1(\mathbb{C}P^1 \setminus \{0,1,\infty\}, t_0)$ is generated by three simple loops based at t_0, one around each of the points $\{0,1,\infty\}$, with the relation that an appropriate composition of the three loops is homotopic to a contractible loop.

Given two linearly independent solutions f, g of $HDE(a,b;c;z)$ near t_0, their analytic continuations along any oriented loop C in $\mathbb{C}P^1 \setminus \{0,1,\infty\}$ starting and ending at t_0 will produce two linearly independent solutions $\tilde{f}, \tilde{g}$ near t_0. This process corresponds to an action of the fundamental group $\pi_1(\mathbb{C}P^1 \setminus \{0,1,\infty\}, t_0)$ on the solution space $S(a,b;c;z)$ of $HDE(a,b;c;z)$ which results in a homomorphism

$$\pi_1(\mathbb{C}P^1 \setminus \{0,1,\infty\}, t_0) \to GL(S(a,b;c;z))$$
$$[C] \mapsto M(C),$$

called the *monodromy representation* of $HDE(a,b;c;z)$. It is determined up to conjugation in $GL_2(\mathbb{C})$ depending on where t_0 is located [15, 105]. The mondromy representation is a degree-2 representation whose image $M(a,b;c;z)$ is called the *monodromy group* of $HDE(a,b;c;z)$. The monodromy group $M(a,b;c;z)$ is generated by the images M_0, M_1, and M_∞ of the three loops described above which generate the fundamental group $\pi_1(\mathbb{C}P^1 \setminus \{0,1,\infty\}, t_0)$, respectively. The *projective monodromy group* is defined to be the image of the monodromy group in $PGL_2(\mathbb{C})$.

EXAMPLE 3.3. Assume $c \notin \mathbb{Z}$, fix t_0 to be near $z = 0$, and let C_0 be a simple clockwise-oriented loop based at t_0 which contains only the singularity 0. Let

$$f = {}_2F_1\begin{bmatrix} a & b \\ & c \end{bmatrix}; z$$
$$g = z^{1-c} {}_2F_1\begin{bmatrix} 1+a-c & 1+b-c \\ & 2-c \end{bmatrix}; z.$$

Then

$$M_0 := M(C_0) = \begin{pmatrix} 1 & 0 \\ 0 & e^{2\pi i(1-c)} \end{pmatrix}.$$

In the case $c \in \mathbb{Z}$, M_0 is conjugate to a matrix of the form $\begin{pmatrix} 1 & * \\ 0 & 1 \end{pmatrix}$.

When $a, b, c \in \mathbb{Q}$, the monodromy group, denoted by $M(a, b; c)$, is a *triangle group*, which, using the standard notation set by Takeuchi in [**91**], is of the form

(3.11) $$(e_1, e_2, e_3) := \langle x, y \mid x^{e_1} = y^{e_2} = (xy)^{e_3} = 1 \rangle,$$

where $e_i \in \mathbb{N} \cup \{\infty\}$ can be computed from a, b, c, and by symmetry we can assume that $e_1 \leq e_2 \leq e_3$. Here $x^\infty = 1$ means the order of x is infinite.

Suppose $c, a+b-c, a-b \notin \mathbb{Z}$. The monodromy group $M(a, b; c; z)$ is $GL_2(\mathbb{C})$-conjugate to the triangle group generated by

$$M_0 = \begin{pmatrix} 1 & 0 \\ 0 & e^{2\pi i(1-c)} \end{pmatrix},$$

$$M_1 = A \begin{pmatrix} 1 & 0 \\ 0 & e^{2\pi i(c-a-b)} \end{pmatrix} A^{-1},$$

$$M_\infty = B \begin{pmatrix} e^{2\pi i a} & 0 \\ 0 & e^{2\pi i b} \end{pmatrix} B^{-1},$$

where

$$A = \begin{pmatrix} \frac{\Gamma(c)\Gamma(c-a-b)}{\Gamma(c-a)\Gamma(c-b)} & \frac{\Gamma(c)\Gamma(a+b-c)}{\Gamma(a)\Gamma(b)} \\ \frac{\Gamma(2-c)\Gamma(c-a-b)}{\Gamma(1-a)\Gamma(1-b)} & \frac{\Gamma(2-c)\Gamma(a+b-c)}{\Gamma(a-c+1)\Gamma(b-c+1)} \end{pmatrix},$$

$$B = \begin{pmatrix} \frac{\Gamma(c)\Gamma(b-a)}{\Gamma(c-a)\Gamma(b)} & \frac{\Gamma(c)\Gamma(a-b)}{\Gamma(c-b)\Gamma(a)} \\ \frac{(-1)^{c-1}\Gamma(2-c)\Gamma(b-a)}{\Gamma(1-a)\Gamma(b-c+1)} & \frac{(-1)^{c-1}\Gamma(2-c)\Gamma(a-b)}{\Gamma(a-c+1)\Gamma(1-b)} \end{pmatrix},$$

which are computed from (3.10) and the Kummer relation

(3.12) $$\,_2F_1 \begin{bmatrix} a & b \\ & c \end{bmatrix}; z \end{bmatrix} = \frac{\Gamma(c)\Gamma(c-a-b)}{\Gamma(c-a)\Gamma(c-b)} \,_2F_1 \begin{bmatrix} a & b \\ & a+b+1-c \end{bmatrix}; 1-z \end{bmatrix}$$
$$+ \frac{\Gamma(c)\Gamma(a+b-c)}{\Gamma(a)\Gamma(b)} (1-z)^{c-a-b} \,_2F_1 \begin{bmatrix} c-a & c-b \\ & 1+c-a-b \end{bmatrix}; 1-z \end{bmatrix},$$

(see Corollary 2.3.3 of [**6**]).

EXAMPLE 3.4. Let $a = 1/4$, $b = 3/4$, $c = 1/2$. Then the generators of the monodromy group $M(\frac{1}{4}, \frac{3}{4}, \frac{1}{2}; z)$ can be computed using the above formulas. We simplify them into the following expressions using the reflection formula (Theorem 2.4) and the multiplication formulas (Theorem 2.6) for the gamma function:

$$M_0 = \begin{pmatrix} 1 & 0 \\ 0 & -1 \end{pmatrix}, \quad M_1 = \begin{pmatrix} 0 & -1/2 \\ -2 & 0 \end{pmatrix}, \quad M_\infty = \begin{pmatrix} 0 & -1/2 \\ 2 & 0 \end{pmatrix},$$

which are each of finite order and satisfy

$$M_0 M_\infty = M_1^{-1} (= M_1) = M_\infty^{-1} M_0.$$

Thus the monodromy group is isomorphic to the triangle group $(2, 2, 4)$, which is also isomorphic to the Dihedral group of order 8. Modulo the center, the projective monodromy group is isomorphic to the triangle group $(2, 2, 2)$, which is isomorphic to the Klein 4-group.

EXAMPLE 3.5. We work out the cases $(a, b, c) = (1/6, 1/3, 5/6)$ and $(1/12, 1/4, 5/6)$.

3.2. SOME PROPERTIES OF HYPERGEOMETRIC FUNCTIONS WITH $n=1$

Let $a = 1/6$, $b = 1/3$, $c = 5/6$. The generators of the monodromy group $M(\frac{1}{6}, \frac{1}{3}, \frac{5}{6}; z)$ computed using the above formulas are

$$M_0 = \begin{pmatrix} 1 & 0 \\ 0 & \zeta_6 \end{pmatrix}, \quad M_1 = A \begin{pmatrix} 1 & 0 \\ 0 & \zeta_3 \end{pmatrix} A^{-1}, \quad M_\infty = B \begin{pmatrix} \zeta_6 & 0 \\ 0 & \zeta_3 \end{pmatrix} B^{-1},$$

where matrices A and B can be computed from the formulas above and simplified to the following expressions using the functional equations of the gamma function

$$A = \begin{pmatrix} \frac{\Gamma(\frac{5}{6})\Gamma(\frac{1}{3})}{\Gamma(\frac{1}{2})\Gamma(\frac{2}{3})} & -3\frac{\Gamma(\frac{5}{6})\Gamma(\frac{2}{3})}{\Gamma(\frac{1}{6})\Gamma(\frac{1}{3})} \\ \frac{1}{6}\frac{\Gamma(\frac{1}{6})\Gamma(\frac{1}{3})}{\Gamma(\frac{5}{6})\Gamma(\frac{2}{3})} & -\frac{1}{2}\frac{\Gamma(\frac{1}{6})\Gamma(\frac{2}{3})}{\Gamma(\frac{1}{2})\Gamma(\frac{1}{3})} \end{pmatrix}, \quad B = \begin{pmatrix} \sqrt{3} & -6\frac{\Gamma(\frac{5}{6})^2}{\Gamma(\frac{1}{6})\Gamma(\frac{1}{2})} \\ \frac{(-1)^{-1/6}}{6}\frac{\Gamma(\frac{1}{6})^2}{\Gamma(\frac{5}{6})\Gamma(\frac{1}{2})} & (-1)^{5/6}\sqrt{3} \end{pmatrix}.$$

Thus both the monodromy group and the projective monodromy group are isomorphic to the triangle group $(3, 6, 6)$.

Let $a = 1/12$, $b = 1/4$, $c = 5/6$. The monodromy group $M(\frac{1}{12}, \frac{1}{4}, \frac{5}{6}; z)$ are generated by

$$M_0 = \begin{pmatrix} 1 & 0 \\ 0 & \zeta_6 \end{pmatrix}, \quad M_1 = A \begin{pmatrix} 1 & 0 \\ 0 & -1 \end{pmatrix} A^{-1}, \quad M_\infty = B \begin{pmatrix} \zeta_{12} & 0 \\ 0 & \sqrt{-1} \end{pmatrix} B^{-1},$$

where

$$A = \begin{pmatrix} \frac{\Gamma(\frac{5}{6})\Gamma(\frac{1}{2})}{\Gamma(\frac{3}{4})\Gamma(\frac{7}{12})} & -2\frac{\Gamma(\frac{5}{6})\Gamma(\frac{1}{2})}{\Gamma(\frac{1}{12})\Gamma(\frac{1}{4})} \\ \frac{1}{6}\frac{\Gamma(\frac{1}{6})\Gamma(\frac{1}{2})}{\Gamma(\frac{11}{12})\Gamma(\frac{3}{4})} & -\frac{1}{3}\frac{\Gamma(\frac{1}{6})\Gamma(\frac{1}{2})}{\Gamma(\frac{1}{4})\Gamma(\frac{5}{12})} \end{pmatrix} = \Gamma\left(\frac{1}{2}\right) \begin{pmatrix} \frac{\Gamma(\frac{5}{6})}{\Gamma(\frac{3}{4})\Gamma(\frac{7}{12})} & -2\frac{\Gamma(\frac{5}{6})}{\Gamma(\frac{1}{12})\Gamma(\frac{1}{4})} \\ \frac{1}{6}\frac{\Gamma(\frac{1}{6})}{\Gamma(\frac{11}{12})\Gamma(\frac{3}{4})} & -\frac{1}{3}\frac{\Gamma(\frac{1}{6})}{\Gamma(\frac{1}{4})\Gamma(\frac{5}{12})} \end{pmatrix},$$

and $B = \begin{pmatrix} \sqrt{2} & -6\frac{\Gamma(\frac{5}{6})^2}{\Gamma(\frac{1}{12})\Gamma(\frac{7}{12})} \\ \frac{(-1)^{-1/6}}{6}\frac{\Gamma(\frac{1}{6})^2}{\Gamma(\frac{5}{12})\Gamma(\frac{11}{12})} & (-1)^{5/6}\sqrt{2} \end{pmatrix}.$

Thus the monodromy group is isomorphic to the triangle group $(2, 6, 12)$. Modulo the center, the projective monodromy group is isomorphic to the triangle group $(2, 6, 6)$.

We note that $S_\Delta(p, q, r)$ (see §3.2.2) is not the same as the monodromy group written in the notation of the triangle group (e_1, e_2, e_3). The Schwarz map is defined as f/g, so $\Delta(p, q, r)$ is related to the projective monodromy group. In fact, the projective monodromy group of $HDE(a, b; c; z)$ is isomorphic to the subgroup of index 2 of the group $S_\Delta(p, q, r)$, consisting of all elements which are products of an even number of reflections.

3.2.5. Schwarz's list and algebraic hypergeometric identities. In [83], Schwarz determined a list of 15 (unordered) triples

$$\{|1 - c|, |c - a - b|, |a - b|\},$$

for which the corresponding differential equations $HDE(a, b; c; z)$ have finite monodromy groups. Beukers and Heckman [17, Theorem 4.8] have generalized Schwarz's work by finding general $_{n+1}F_n$ with rational parameters and finite monodromy groups. It is well-known that

$$_2F_1 \begin{bmatrix} a & b \\ & c \end{bmatrix}; z$$

is an algebraic function of z when the monodromy group for $HDE(a, b; c; z)$ is a finite group; we give an example of this phenomenon below.

EXAMPLE 3.6. For a classical example, let $r = \frac{i}{N}$ be reduced, with $\frac{N}{2} < i < N$. Then, to compute the Schwarz triangle corresponding to $HDE(r - \frac{1}{2}, r; 2r; z)$, we see that $p = |1 - 2r|$, $q = \frac{1}{2}$, $r = \frac{1}{2}$. Thus the corresponding triangle group is $(2, 2, 1/|1 - 2r|)$, which can be identified with a Dihedral group. This corresponds to a tiling of the unit sphere, since $\frac{1}{2} + \frac{1}{2} + |1 - 2r| > 1$. The corresponding algebraic expression for the hypergeometric functions is equation (1.3), as we have seen before.

In particular, when $N = 3$, $i = 2$, i.e. when $r = \frac{2}{3}$, by the algorithm described above, the monodromy group $M(\frac{1}{6}, \frac{2}{3}; \frac{4}{3})$ is generated by

$$M_0 = \begin{pmatrix} 1 & 0 \\ 0 & \zeta_3^2 \end{pmatrix}, \quad M_1 = \begin{pmatrix} 0 & 2^{2/3} \\ 2^{-2/3} & 0 \end{pmatrix}, \quad M_\infty = \begin{pmatrix} 0 & 2^{2/3} \\ 2^{-2/3}\zeta_3 & 0 \end{pmatrix},$$

which is isomorphic to the triangle group $(2, 3, 6)$. To be more precise, we have

$$A = \begin{pmatrix} \frac{\Gamma(\frac{4}{3})\Gamma(\frac{1}{2})}{\Gamma(\frac{7}{6})\Gamma(\frac{2}{3})} & \frac{\Gamma(\frac{4}{3})\Gamma(-\frac{1}{2})}{\Gamma(\frac{1}{6})\Gamma(\frac{2}{3})} \\ \frac{\Gamma(\frac{2}{3})\Gamma(\frac{1}{2})}{\Gamma(\frac{5}{6})\Gamma(\frac{1}{3})} & \frac{\Gamma(\frac{2}{3})\Gamma(-\frac{1}{2})}{\Gamma(-\frac{1}{6})\Gamma(\frac{1}{3})} \end{pmatrix}, \quad B = \begin{pmatrix} \frac{\Gamma(\frac{4}{3})\Gamma(\frac{1}{2})}{\Gamma(\frac{7}{6})\Gamma(\frac{2}{3})} & \frac{\Gamma(\frac{4}{3})\Gamma(-\frac{1}{2})}{\Gamma(\frac{1}{6})\Gamma(\frac{2}{3})} \\ (-1)^{1/3}\frac{\Gamma(\frac{2}{3})\Gamma(\frac{1}{2})}{\Gamma(\frac{5}{6})\Gamma(\frac{1}{3})} & (-1)^{1/3}\frac{\Gamma(\frac{2}{3})\Gamma(-\frac{1}{2})}{\Gamma(-\frac{1}{6})\Gamma(\frac{1}{3})} \end{pmatrix}$$

and using the identities

$$\Gamma(1 + z) = z\Gamma(z), \quad \Gamma(z)\Gamma(1 - z) = \pi/\sin \pi z, \quad \Gamma(1/2) = \sqrt{\pi}$$

and

$$\Gamma\left(\frac{1}{3}\right) = \Gamma\left(2 \cdot \frac{1}{6}\right) = \frac{2^{1/3}}{2\sqrt{\pi}}\Gamma\left(\frac{1}{6}\right)\Gamma\left(\frac{2}{3}\right),$$

we obtain

$$A = \begin{pmatrix} 2^{1/3} & -\frac{1}{3}2^{1/3} \\ 2^{-1/3} & \frac{1}{3}2^{-1/3} \end{pmatrix}, \quad B = \begin{pmatrix} 2^{1/3} & -\frac{1}{3}2^{1/3} \\ \zeta_6 2^{-1/3} & \zeta_6 \frac{1}{3}2^{-1/3} \end{pmatrix}$$

and thus

$$M_1 = A \begin{pmatrix} 1 & 0 \\ 0 & -1 \end{pmatrix} A^{-1} = \begin{pmatrix} 0 & 2^{2/3} \\ 2^{-2/3} & 0 \end{pmatrix},$$

$$M_\infty = B \begin{pmatrix} \zeta_6 & 0 \\ 0 & \zeta_3^2 \end{pmatrix} B^{-1} = \begin{pmatrix} 0 & 2^{2/3} \\ 2^{-2/3}\zeta_3 & 0 \end{pmatrix}.$$

The projective monodromy group is isomorphic to the triangle group $(2, 2, 3)$, which can be identified with the Dihedral group of size 6. This corresponds to the tiling of the sphere depicted below.

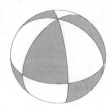

3.2. SOME PROPERTIES OF HYPERGEOMETRIC FUNCTIONS WITH $n=1$

In the above picture, one can think of each darker (respectively lighter) triangle as the image of the upper (respectively lower) half complex plane under the corresponding Schwarz map as described in Theorem 3.2.

Furthermore, when $N=8$, $i=5$, the monodromy group $M(\frac{1}{8},\frac{5}{8};\frac{5}{4};z)$ is $(2,4,8)$ and the projective monodromy group is the triangle group $(2,2,4)$, which can be identified with the Dihedral group of size 8.

EXAMPLE 3.7. When $a=\frac{1}{4}, b=\frac{3}{4}, c=\frac{4}{3}$ or $\frac{2}{3}$ the corresponding Schwarz triangle is $(\pi/2, \pi/3, \pi/3)$ and the monodromy group is generated by

$$M_0 = \begin{pmatrix} 1 & 0 \\ 0 & \zeta_3^2 \end{pmatrix}, \quad M_1 = \begin{pmatrix} \frac{\sqrt{-3}}{3} & \frac{\Gamma(\frac{1}{3})^2}{\Gamma(\frac{1}{12})\Gamma(\frac{7}{12})}\frac{8\sqrt{2}}{\sqrt{3}+i} \\ \frac{\sqrt{3}-i}{12\sqrt{2}}\frac{\Gamma(\frac{1}{12})\Gamma(\frac{7}{12})}{\Gamma(\frac{1}{3})^2} & \frac{3+\sqrt{-3}}{6} \end{pmatrix},$$

$$M_\infty = \begin{pmatrix} -\frac{\sqrt{-3}}{3} & \frac{\Gamma(\frac{1}{3})^2}{\Gamma(\frac{1}{12})\Gamma(\frac{7}{12})}\frac{8\sqrt{2}}{\sqrt{3}-i} \\ \frac{\Gamma(\frac{2}{3})^2}{\Gamma(\frac{5}{12})\Gamma(\frac{11}{12})}\frac{i-\sqrt{3}}{4\sqrt{2}} & \frac{\sqrt{-3}}{3} \end{pmatrix},$$

with $M_0^3 = M_1^3 = id$, $M_\infty^2 = -id$, $M_\infty M_1 M_0 = id$. This is isomorphic to the triangle group $(3,3,4)$ and the projective monodromy group is isomorphic to $(2,3,3)$. In a private communication, Yifan Yang told us the following, which he computed numerically. Let

$$f(t) := {}_2F_1\left[\begin{matrix} \frac{1}{4} & \frac{3}{4} \\ & \frac{4}{3} \end{matrix}; t\right].$$

Then

$$4096 - 4096 f^2 + 1152 t f^4 - 27 t^2 f^8 = 0.$$

For a generic choice of t, the Galois group of the splitting field of the polynomial $4096 - 4096 x + 1152 t x^2 - 27 t^2 x^4$ over $\mathbb{Q}(t)$ is isomorphic to the symmetric group S_4. For special choices of t, the Galois group is smaller. For instance when $t=1/2$, the Galois group is isomorphic to D_4, the Dihedral group of order 8.

Similarly, Yifan Yang computed that

$$g(t) := {}_2F_1\left[\begin{matrix} \frac{1}{4} & \frac{3}{4} \\ & \frac{2}{3} \end{matrix}; t\right]$$

satisfies

$$1 + 8g^2 + 18(1-t)g^4 - 27(1-t)^2 g^8 = 0.$$

We have verified his claims by using the recursive relations satisfied by the coefficients of f and g. See §8.3, Example 8.16, specifically (8.9), for the finite field analogues of $f(t)^2$ and $g(t)^2$ with $t = \frac{\lambda}{\lambda-1}$.

Yang's first result is related, by a Pfaff transformation, to the first formula on page 165 of [**98**] of Vidūnas:

$${}_2F_1\left[\begin{matrix} \frac{1}{4} & -\frac{1}{12} \\ & \frac{2}{3} \end{matrix}; \frac{x(x+4)^3}{4(2x-1)^3}\right] = \frac{1}{(1-2x)^{1/4}}.$$

We give a numeric finite field analogue of this formula in Conjecture 9.25.

We note here that Vidūnas [**98**] gave a list of several other algebraic hypergeometric ${}_2F_1$ functions. Moreover, Baldassarri and Dwork [**12**] determined general second order linear differential equations with only algebraic solutions.

3.2.6. Evaluation Formulas. There are many evaluation and transformation formulas for the classical $_2F_1$ hypergeometric functions that are incredibly useful for obtaining higher transformation formulas, computing periods, or proving supercongruences. For example, this technique is employed in work of the second, third, and fourth authors [66, 68, 90], among others. These evaluation formulas are obtained in various ways; for example, see [6, 38, 88, 102].

First, the Gauss summation formula [6, Thm. 2.2.2] can be obtained by using the integral form of the $_2F_1$ from (3.4) and (3.5), together with the beta integral from Definition 2.2. In particular when $\operatorname{Re}(c - a - b) > 0$,

$$(3.13) \qquad {}_2F_1\begin{bmatrix} a & b \\ & c \end{bmatrix}; 1\end{bmatrix} = \frac{B(b, c - a - b)}{B(b, c - b)} = \frac{\Gamma(c)\Gamma(c - a - b)}{\Gamma(c - a)\Gamma(c - b)}.$$

Next, we have the Kummer evaluation theorem [6, Cor. 3.1.2], which gives that

$$(3.14) \qquad {}_2F_1\begin{bmatrix} c & b \\ & 1 + c - b \end{bmatrix}; -1\end{bmatrix} = \frac{B(1 + c - b, \frac{c}{2} + 1)}{B(1 + c, \frac{c}{2} + 1 - b)}.$$

Also, we have the very useful Pfaff-Saalschütz formula [6, Thm. 2.2.6]. This result follows from comparing coefficients on both sides of (3.9). It states that for a positive integer n and $a, b, c \in \mathbb{C}$,

$$(3.15) \quad {}_3F_2\begin{bmatrix} a & b & -n \\ c & 1 + a + b - n - c \end{bmatrix}; 1\end{bmatrix} = \frac{(c - a)_n (c - b)_n}{(c)_n (c - a - b)_n}$$

$$= \frac{\Gamma(c - a + n)\Gamma(c - b + n)\Gamma(c)\Gamma(c - a - b)}{\Gamma(c - a)\Gamma(c - b)\Gamma(c + n)\Gamma(c - a - b + n)}$$

$$= \frac{B(c - a + n, c - b + n)B(c, c - a - b)}{B(c + n, c - a - b + n)B(c - a, c - b)}.$$

Note that the left side is a terminating series as $-n$ is a negative integer while the right side is a product. Such formulas are useful for proving supercongruences as in [68, 90, 108].

In a more symmetric form, it can be written as the following when one of a, b, d is a negative integer,

$$(3.16) \quad {}_3F_2\begin{bmatrix} a & b & d \\ c & 1 + a + b + d - c \end{bmatrix}; 1\end{bmatrix} =$$

$$\frac{\Gamma(c - a - d)\Gamma(c - b - d)\Gamma(c - a - b)\Gamma(c)}{\Gamma(c - a)\Gamma(c - b)\Gamma(c - d)\Gamma(c - a - b - d)}.$$

There is a rich supply of evaluation formulas in the literature, see for instance [6, 38, 102], [88, Appendix III].

3.2.7. Higher transformation formulas. There are also many quadratic or higher order transformation formulas for $_2F_1$ (and a few for $_3F_2$ functions) which are obtained in various ways [40, 94, 97]. An example is the following Kummer quadratic transformation formula [6, Thm. 3.1.1] or [88, 2.3.2.1] which gives that

$${}_2F_1\begin{bmatrix} c & b \\ & c - b + 1 \end{bmatrix}; z\end{bmatrix} = (1 - z)^{-c} \, {}_2F_1\begin{bmatrix} \frac{c}{2} & \frac{1+c}{2} - b \\ & c - b + 1 \end{bmatrix}; \frac{-4z}{(1 - z)^2}\end{bmatrix}.$$

3.2. SOME PROPERTIES OF HYPERGEOMETRIC FUNCTIONS WITH $n = 1$

From Theorem 3.2, we observe that the Schwarz triangle corresponding to the hypergeometric function on the left is $\Delta(|b-c|, |b-c|, |1-2b|)$. It can be tiled by two copies of the Schwarz triangle $\Delta(|b-c|, \frac{1}{2}, \frac{1}{2}|1-2b|)$, which corresponds to the hypergeometric function appearing on the right hand side. In other words, the projective monodromy group corresponding to the left hand side is commensurable with the projective monodromy group on the right hand side. This is illustrated in the diagram below.

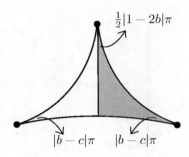

CHAPTER 4

Finite Field Analogues

In this chapter we build on the discussion of period and hypergeometric functions for the classical complex setting described in Chapter 3 by defining period and hypergeometric functions over finite fields which behave analogously. We conclude by comparing our definition to those previously defined by others.

4.1. Periods in the finite field setting

For any function $f : \mathbb{F}_q \longrightarrow \mathbb{C}$, the orthogonality of characters [14, 46] or Greene's Lagrange inversion formula over finite fields (Theorem 2.10 with $g(x) = x$) implies that f has a unique representation

$$(4.1) \qquad f(x) = \delta(x)f(0) + \sum_{\chi \in \widehat{\mathbb{F}_q^\times}} f_\chi \chi(x),$$

where

$$f_\chi = \frac{1}{q-1} \sum_{y \in \mathbb{F}_q} f(y)\overline{\chi}(y).$$

Later, we will use the uniqueness in this statement to obtain identities.

It is also useful to note that when $a \neq 0$,

$$(4.2) \qquad \delta(x - a) = \sum_{\chi \in \widehat{\mathbb{F}_q^\times}} \frac{\chi(x/a)}{q-1}.$$

We now define a new version of hypergeometric functions for the finite field $\mathbb{F}_q$ parallel to §3.1. We start with a natural analogue to (3.3) by letting

$$(4.3) \qquad {}_1\mathbb{P}_0[A; x; q] := \overline{A}(1-x),$$

and then inductively defining

$$(4.4) \quad {}_{n+1}\mathbb{P}_n \begin{bmatrix} A_1 & A_2 & \cdots & A_{n+1} \\ & B_1 & \cdots & B_n \end{bmatrix}; \lambda; q \Bigg] :=$$

$$\sum_{y \in \mathbb{F}_q} A_{n+1}(y)\overline{A}_{n+1}B_n(1-y) \cdot {}_n\mathbb{P}_{n-1}\begin{bmatrix} A_1 & A_2 & \cdots & A_n \\ & B_1 & \cdots & B_{n-1} \end{bmatrix}; \lambda y; q \Bigg],$$

which corresponds to (3.6) via the dictionary in §2.4 (and recalling Remark 2.12).

We note the asymmetry among the A_i (resp. B_j) in the definition. Part of the reason we start with the analogue of the period, rather than the hypergeometric function is because the periods are related to point counting. When there is no ambiguity in our choice of field $\mathbb{F}_q$, we will leave out the q in this notation and simply write

$$_{n+1}\mathbb{P}_n \begin{bmatrix} A_1 & A_2 & \cdots & A_{n+1} \\ & B_1 & \cdots & B_n \end{bmatrix};\lambda \Bigg] := {}_{n+1}\mathbb{P}_n \begin{bmatrix} A_1 & A_2 & \cdots & A_{n+1} \\ & B_1 & \cdots & B_n \end{bmatrix};\lambda;q \Bigg].$$

When $n = 1$ we have,

(4.5) $$_2\mathbb{P}_1 \begin{bmatrix} A_1 & A_2 \\ & B_1 \end{bmatrix};\lambda \Bigg] = \sum_{y \in \mathbb{F}_q} A_2(y)\overline{A_2}B_1(1-y)\overline{A_1}(1-\lambda y).$$

In terms of Jacobi sums,

(4.6) $$_2\mathbb{P}_1 \begin{bmatrix} A_1 & A_2 \\ & B_1 \end{bmatrix};\lambda \Bigg] = \begin{cases} J(A_2, \overline{A_2}B_1), & \text{if } \lambda = 0 \\ \dfrac{1}{q-1} \displaystyle\sum_{\chi \in \widehat{\mathbb{F}_q^\times}} J(\overline{A_1},\overline{\chi})J(A_2\chi, B_1\overline{A_2})\chi(\lambda), & \text{if } \lambda \neq 0 \end{cases}.$$

By Lemma 2.8
(4.7)
$$_2\mathbb{P}_1 \begin{bmatrix} A_1 & A_2 \\ & B_1 \end{bmatrix};\lambda \Bigg] = \begin{cases} J(A_2, \overline{A_2}B_1), & \text{if } \lambda = 0, \\ \dfrac{A_2(-1)}{q-1} \displaystyle\sum_{\chi \in \widehat{\mathbb{F}_q^\times}} J(A_1\chi, \overline{\chi})J(A_2\chi, \overline{B_1\chi})\chi(\lambda), & \text{if } \lambda \neq 0. \end{cases}$$

While they are equivalent, the second expression is more symmetric when the finite field analogue of binomial coefficients defined in (2.18) is used. A general formula for the $_{n+1}\mathbb{P}_n$ period function highlighting this symmetry is:

$$_{n+1}\mathbb{P}_n \begin{bmatrix} A_1 & A_2 & \cdots & A_{n+1} \\ & B_1 & \cdots & B_n \end{bmatrix};\lambda \Bigg]$$

$$= \frac{(-1)^{n+1}}{q-1} \cdot \left(\prod_{i=1}^n A_{i+1}B_i(-1)\right) \sum_{\chi \in \widehat{\mathbb{F}_q^\times}} \binom{A_1\chi}{\chi}\binom{A_2\chi}{B_1\chi}\cdots \binom{A_{n+1}\chi}{B_n\chi}\chi(\lambda)$$

$$+ \delta(\lambda) \prod_{i=1}^n J(A_{i+1}, \overline{A_{i+1}}B_i).$$

This is similar to Greene's definition [**41**, Def. 3.10], though we note that our binomial coefficient differs from Greene's by a factor of $-q$, and we define the value at $\lambda = 0$ differently. Our version gives a natural reference point for the normalization to come later, as it does in the classical case.

4.2. Hypergeometric varieties

It is well-known that Greene's hypergeometric functions can be used to count points on varieties over finite fields ([**4**], [**13**], [**23**], [**24**], [**35**], [**37**], [**51**], [**56**], [**57**], [**75**], [**81**], [**82**], [**96**], et al.) and evaluate twisted Kloosterman sums [**61**]. For example, the following result of Ono (stated here in terms of our notation) gives the number of points on the Legendre curves $E_\lambda : y^2 = x(x-1)(x-\lambda)$ over $\mathbb{F}_q$.

THEOREM 4.1 (Ono [75]). *For $\lambda \in \mathbb{F}_q$ with $\lambda \neq 0, 1$,*

$$\#E_\lambda(\mathbb{F}_q) = q + 1 + {}_2\mathbb{P}_1 \begin{bmatrix} \phi & \phi \\ & \varepsilon \end{bmatrix}; \lambda \end{bmatrix}.$$

See also [81] by Rouse.

We use the period functions ${}_{n+1}\mathbb{P}_n$ for this purpose in a more general context. Consider the family of *hypergeometric algebraic varieties*

$$X_\lambda: \quad y^N = x_1^{i_1} \cdots x_n^{i_n} \cdot (1-x_1)^{j_1} \cdots (1-x_n)^{j_n} \cdot (1-\lambda x_1 \cdots x_n)^k.$$

We can count the points on X_λ in a manner similar to that in [24], where the case $N = n+1$, $i_s = n$, $j_t = 1$ for all s, t and $k = 1$ is considered. Different algebraic models are used for the hypergeometric algebraic varieties over $\mathbb{Q}$ in [16, Thm. 1.5]. We choose our model in order to make the next statement straightforward.

PROPOSITION 4.2. *Let $q = p^e \equiv 1 \pmod{N}$ be a prime power, and $\eta_N \in \widehat{\mathbb{F}_q^\times}$ a primitive order N character. Then*

$$\#X_\lambda(\mathbb{F}_q) = 1 + q^n + \sum_{m=1}^{N-1} {}_{n+1}\mathbb{P}_n \begin{bmatrix} \eta_N^{-mk} & \eta_N^{mi_n} & \cdots & \eta_N^{mi_1} \\ & \eta_N^{mi_n+mj_n} & \cdots & \eta_N^{mi_1+mj_1} \end{bmatrix}; \lambda; q \end{bmatrix}.$$

PROOF. We begin as in the proof of Theorem 2 in [24], to see that

$$\#X_\lambda(\mathbb{F}_q) = 1 + q^n$$
$$+ \sum_{m=1}^{N-1} \sum_{x_i \in \mathbb{F}_q} \eta_N^m(x_1^{i_1} \cdots x_n^{i_n}(1-x_1)^{j_1} \cdots (1-x_n)^{j_n}(1-\lambda x_1 \cdots x_n)^k).$$

Then, by applying (4.4) n times, we see that

$${}_{n+1}\mathbb{P}_n \begin{bmatrix} \eta_N^{-mk} & \eta_N^{mi_n} & \cdots & \eta_N^{mi_1} \\ & \eta_N^{mi_n+mj_n} & \cdots & \eta_N^{mi_1+mj_1} \end{bmatrix}; \lambda; q \end{bmatrix}$$
$$= \sum_{x_i \in \mathbb{F}_q} \eta_N^{mi_1}(x_1)\eta_N^{mj_1}(1-x_1) \cdots \eta_N^{mi_n}(x_n)\eta_N^{mj_n}(1-x_n)\eta_N^{mk}(1-\lambda x_1 \cdots x_n).$$

This gives the result. □

It is more convenient to obtain the major term $1 + q^n$ by using $\varepsilon(0) = 1$. The total point count is independent of the choice of $\varepsilon(0)$. Also, we are not dealing with the smooth model of X_λ here. The point count here is related to Dwork's work on zeta functions of hypersurfaces which will be recalled in §6.6.

4.3. Hypergeometric functions over finite fields

In the classical case (see §3.1), a hypergeometric function is normalized to have constant term 1, obtained from the corresponding ${}_{n+1}P_n$ function divided by its value at 0. Here we will similarly normalize the finite field period functions. We thus define

$$(4.8) \qquad {}_2\mathbb{F}_1 \begin{bmatrix} A_1 & A_2 \\ & B_1 \end{bmatrix}; \lambda \end{bmatrix} := \frac{1}{J(A_2, B_1\overline{A_2})} \cdot {}_2\mathbb{P}_1 \begin{bmatrix} A_1 & A_2 \\ & B_1 \end{bmatrix}; \lambda \end{bmatrix}.$$

The ${}_2\mathbb{F}_1$ function satisfies

1) $_2\mathbb{F}_1\begin{bmatrix} A_1 & A_2 \\ & B_1 \end{bmatrix};0\end{bmatrix} = 1;$

2) $_2\mathbb{F}_1\begin{bmatrix} A_1 & A_2 \\ & B_1 \end{bmatrix};\lambda\end{bmatrix} = {}_2\mathbb{F}_1\begin{bmatrix} A_2 & A_1 \\ & B_1 \end{bmatrix};\lambda\end{bmatrix}$, if $A_1, A_2 \neq \varepsilon$, and $A_1, A_2 \neq B_1$.

Property 1) follows from the definition and (4.5); 2) will be proved in Proposition 8.7 below. Intuitively, with the additional Jacobi sum factor one can rewrite the right hand side of (4.8) using the finite field analogues of rising factorials with the roles of A_1 and A_2 being symmetric.

More generally, we define

(4.9) $\quad _{n+1}\mathbb{F}_n\begin{bmatrix} A_1 & A_2 & \cdots & A_{n+1} \\ & B_1 & \cdots & B_n \end{bmatrix};\lambda\end{bmatrix}$

$$:= \frac{1}{\prod_{i=1}^n J(A_{i+1}, B_i\overline{A_{i+1}})} {}_{n+1}\mathbb{P}_n\begin{bmatrix} A_1 & A_2 & \cdots & A_{n+1} \\ & B_1 & \cdots & B_n \end{bmatrix};\lambda\end{bmatrix}.$$

DEFINITION 4.3. A period function $_{n+1}\mathbb{P}_n\begin{bmatrix} A_1 & A_2 & \cdots & A_{n+1} \\ & B_1 & \cdots & B_n \end{bmatrix};\lambda\end{bmatrix}$ or the corresponding $_{n+1}\mathbb{F}_n\begin{bmatrix} A_1 & A_2 & \cdots & A_{n+1} \\ & B_1 & \cdots & B_n \end{bmatrix};\lambda\end{bmatrix}$ is said to be *primitive* if $A_i \neq \varepsilon$ and $A_i \neq B_j$ for all i, j; otherwise it is said to be *imprimitive*. Similarly, we call the corresponding classical period or hypergeometric function *primitive* if $a_i, a_i - b_j \notin \mathbb{Z}$ for any a_i, b_j.

As in the $n = 1$ case, the value of the $_{n+1}\mathbb{F}_n$ functions at $\lambda = 0$ is 1, and for primitive $_{n+1}\mathbb{F}_n$ the characters A_i (resp. B_j) can be permuted without effect, which can be seen from (4.10) below.

4.4. Comparison with other finite field hypergeometric functions

Alternative definitions for hypergeometric functions over finite fields have been given in the papers of Greene [41], McCarthy [70], Katz [47, Ch. 8.2], and Beukers, Cohen, Mellit [16]. Sometimes, they are referred to as Gaussian hypergeometric functions or finite hypergeometric functions. For the sake of consistency within this paper, below we use the notation $_j\mathbb{F}_k$ rather than $_jF_k$ which is used in [41] and [70]. Greene's version is defined by

$$_{n+1}\mathbb{F}_n\begin{pmatrix} A_1 & A_2 & \cdots & A_{n+1} \\ & B_1 & \cdots & B_n \end{pmatrix};\lambda\end{pmatrix}^G :=$$

$$\frac{(-1)^{n+1}}{q^n(q-1)} \cdot \sum_{\chi \in \widehat{\mathbb{F}_q^\times}} \binom{A_1\chi}{\chi}\binom{A_2\chi}{B_1\chi}\cdots\binom{A_{n+1}\chi}{B_n\chi}\chi(\lambda).$$

McCarthy's version is defined by

$$_{n+1}\mathbb{F}_n\begin{pmatrix} A_1 & A_2 & \cdots & A_{n+1} \\ & B_1 & \cdots & B_n \end{pmatrix};\lambda\end{pmatrix}^M :=$$

$$\frac{1}{q-1} \sum_{\chi \in \widehat{\mathbb{F}_q^\times}} \prod_{i=1}^{n+1} \frac{g(A_i\chi)}{g(A_i)} \prod_{j=1}^n \frac{g(\overline{B_j}\overline{\chi})}{g(\overline{B_j})} g(\overline{\chi})\chi(-1)^{n+1}\chi(\lambda)$$

4.4. COMPARISON WITH OTHER FINITE FIELD HYPERGEOMETRIC FUNCTIONS

and is symmetric in the A_i's and B_j's. It can be shown that the "hypergeometric sum" defined by Katz in [47] can be written as

$$_n\mathbb{F}_m \begin{pmatrix} A_1 & A_2 & \cdots & A_n \\ B_1 & B_2 & \cdots & B_m \end{pmatrix}; \lambda \end{pmatrix}^K := \frac{1}{q-1} \sum_{\chi \in \widehat{\mathbb{F}_q^\times}} \overline{\chi}(\lambda) \prod_{i=1}^n g(A_i\chi) \prod_{j=1}^m g(\overline{B_j}\chi) B_j \chi(-1).$$

In [16], the authors used the following modified version of Katz's hypergeometric sum for the case $m = n$

$$_m\mathbb{F}_m \begin{pmatrix} A_1 & A_2 & \cdots & A_m \\ B_1 & B_2 & \cdots & B_m \end{pmatrix}; \lambda \end{pmatrix}^{BCM} := \frac{1}{1-q} \sum_{\chi \in \widehat{\mathbb{F}_q^\times}} \chi(\lambda) \prod_{i=1}^m \frac{g(A_i\chi)}{g(A_i)} \frac{g(\overline{B_j}\chi)}{g(\overline{B_j})} \chi(-1),$$

which is equivalent to McCarthy's version with $B_1 = \varepsilon$. In the work of [16], the hypergeometric sum is the finite field version of the hypergeometric functions corresponding to hypergeometric motives over $\mathbb{Q}$, so we have additional conditions on the characters A_i, B_j such that the sets $\{A_i\}$ and $\{B_j\}$ are closed under Galois conjugates respectively.

Our period functions are closely related to Greene's Gaussian hypergeometric functions and both can be used to count points. The relationship between the two is

$$_{n+1}\mathbb{P}_n \begin{bmatrix} A_1 & A_2 & \cdots & A_{n+1} \\ & B_1 & \cdots & B_n \end{bmatrix}; \lambda \end{bmatrix}$$
$$= q^n \left(\prod_{i=1}^n A_{i+1} B_i(-1) \right) {}_{n+1}\mathbb{F}_n \begin{pmatrix} A_1 & A_2 & \cdots & A_{n+1} \\ & B_1 & \cdots & B_n \end{pmatrix}; \lambda \end{pmatrix}^G$$
$$+ \delta(\lambda) \prod_{i=1}^n J(A_{i+1}, \overline{A_{i+1}} B_i).$$

In the primitive case, the normalized $_{n+1}\mathbb{F}_n$-hypergeometric function defined in (4.9) is the same as McCarthy's hypergeometric function over finite fields, when $\lambda \neq 0$. To be precise, in the primitive case

$$(4.10) \quad {}_{n+1}\mathbb{F}_n \begin{bmatrix} A_1 & A_2 & \cdots & A_{n+1} \\ & B_1 & \cdots & B_n \end{bmatrix}; \lambda \end{bmatrix}$$
$$= \frac{1}{\prod_{i=1}^n J(A_{i+1}, B_i \overline{A_{i+1}})} {}_{n+1}\mathbb{P}_n \begin{bmatrix} A_1 & A_2 & \cdots & A_{n+1} \\ & B_1 & \cdots & B_n \end{bmatrix}; \lambda \end{bmatrix}$$
$$= {}_{n+1}\mathbb{F}_n \begin{pmatrix} A_1 & A_2 & \cdots & A_{n+1} \\ & B_1 & \cdots & B_n \end{pmatrix}; \lambda \end{pmatrix}^M + \delta(\lambda).$$

We note that McCarthy's hypergeometric function is related to Katz's via (see Prop. 2.6 of [70])

$$_{n+1}\mathbb{F}_n\begin{pmatrix} A_1 & A_2 & \cdots & A_{n+1} \\ & B_1 & \cdots & B_n \end{pmatrix};\lambda\end{pmatrix}^M = \left[\frac{1}{g(A_1)}\prod_{i=1}^n \frac{B_i(-1)}{g(A_{i+1})g(\overline{B_i})}\right]$$
$$\cdot {}_{n+1}\mathbb{F}_{n+1}\begin{pmatrix} A_1 & A_2 & \cdots & A_{n+1} \\ \varepsilon & B_1 & \cdots & B_n \end{pmatrix};\frac{1}{\lambda}\end{pmatrix}^K.$$

CHAPTER 5

Some Related Topics on Galois Representations

In this chapter, we interpret the finite field analogues of periods and hypergeometric functions using a Galois perspective. See books by Serre [**84**, **85**] on basic representation theory and Galois representations. Readers can choose to skip this chapter as most of the later proofs can be obtained using the setup in the previous chapters. The Galois interpretation gives us a global picture and allows us to predict the finite field analogues of classical formulas quite efficiently. The approach below is derived using work of Weil [**101**] and is used to reinterpret some results in [**23**] on generalized Legendre curves. See [**47**] by Katz and [**16**] by Beukers, Cohen and Mellit for related more general discussions on the topic.

5.1. Absolute Galois groups and Galois representations

We now recall some standard results in algebraic number theory. Let J/L be a finite Galois extension of number fields with Galois group $\mathrm{Gal}(J/L)$, and let $\mathcal{O}_J$ and $\mathcal{O}_L$ be the respective rings of integers. As $\mathcal{O}_J$ and $\mathcal{O}_L$ are Dedekind domains, every prime ideal $\mathfrak{p}$ of $\mathcal{O}_L$ factors in $\mathcal{O}_J$ into a product of prime ideals $\wp_i$ over $\mathfrak{p}$. Since the Galois group $\mathrm{Gal}(J/L)$ acts transitively on primes over $\mathfrak{p}$, this factorization has the form $\prod_{i=1}^{g} \wp_i^e$, where e is called the ramification degree of $\mathfrak{p}$ in L and is independent of $\wp_i$ as J/L is Galois. The transitive action also implies the quantity $f = [\mathcal{O}_J/\wp_i : \mathcal{O}_L/\mathfrak{p}]$ is independent of i. For a prime ideal $\wp_i$ of $\mathcal{O}_J$, the *decomposition group* of $\wp_i$ is defined by $D_{\wp_i} = \{\sigma \in \mathrm{Gal}(J/L) \mid \sigma(\wp_i) = \wp_i\}$. One easily sees from the transitive action that the decomposition groups $D_{\wp_i}$ are conjugate to one another within $\mathrm{Gal}(J/L)$. Thus if $\mathrm{Gal}(J/L)$ is abelian we have that $D_{\wp_i}$ depends only on $\mathfrak{p}$. Furthermore, $\#D_{\wp_i} = ef$, and $efg = [J:L]$.

The discriminant $\mathrm{Dis}_{J/L}$ ideal is a product of powers of prime ideals of $\mathcal{O}_L$. If a prime does not divide $\mathrm{Dis}_{J/L}$ then it is a standard result that $e = 1$. Such primes are called *unramified*. Assume $\mathfrak{p}$ is unramified in J/L. As the extension $(\mathcal{O}_J/\wp_i)/(\mathcal{O}_L/\mathfrak{p})$ is a finite extension of finite fields it has cyclic Galois group with generator given by $x \mapsto x^{\#\mathcal{O}_L/\mathfrak{p}}$. Any element of $Gal(J/L)$ whose restriction to the residue field $\mathcal{O}_J/\wp_i$ is the inverse of the above map is called a (geometric) *Frobenius* automorphism. In this case

$$\mathbb{Z}/f\mathbb{Z} \simeq \mathrm{Gal}\left((\mathcal{O}_J/\wp_i)/(\mathcal{O}_L/\mathfrak{p})\right) \simeq D_{\wp_i} \subset \mathrm{Gal}(J/L).$$

Thus for unramified primes $\mathfrak{p}$ in J/L we have the well-defined *Frobenius conjugacy class*, $\mathrm{Frob}_{\mathfrak{p}} \subset \mathrm{Gal}(J/L)$.

EXAMPLE 5.1. Let ζ_5 be a primitive 5th root of unity and set $J = \mathbb{Q}(\zeta_5)$ and $L = \mathbb{Q}$. Then $\mathrm{Dis}_{J/L} = (5^3)$ and all primes other than (5) are unramified. For an unramified prime p, f is the smallest integer satisfying $p^f \equiv 1 \pmod{5}$. One determines g by solving $efg = [\mathbb{Q}(\zeta_5) : \mathbb{Q}] = 4$. The primes p that are 1 modulo 5 have $(f,g) = (1,4)$. For primes that are 2 and 3 modulo 5 we have $(f,g) = (4,1)$. Finally, primes that are 4 modulo 5 have $(f,g) = (2,2)$.

In this example $\text{Gal}(\mathbb{Q}(\zeta_5)/\mathbb{Q}) \simeq \mathbb{Z}/4\mathbb{Z}$ is cyclic. In a biquadratic extension such as $\mathbb{Q}(\sqrt{5}, \sqrt{13})/\mathbb{Q}$ with Galois group $(\mathbb{Z}/2\mathbb{Z})^2$ one will never have an unramified prime p with Frob_p having order 4. It will either be the case that $(f,g) = (2,2)$ or $(f,g) = (1,4)$.

Arithmetic objects over a number field L often have a representation of the absolute Galois group of L, $G_L := \text{Gal}(\overline{L}/L)$, attached to them. For instance given an elliptic curve $E_{/L}$ one can study its m-torsion $E[m]$ over $\overline{L}$. Since $E(\mathbb{C}) \simeq S^1 \times S^1$ we see $E[m] \simeq \mathbb{Z}/m\mathbb{Z} \times \mathbb{Z}/m\mathbb{Z}$. More importantly, the abelian group structure of $E(\overline{L}) \subset E(\mathbb{C})$ is defined over L so $E(\overline{L})$ comes equipped with an action of G_L. Thus we have a continuous homomorphism $G_L \to \text{Aut}(E[m]) \subset GL_2(\mathbb{Z}/m\mathbb{Z})$. Now fix a prime ℓ, set $m = \ell^n$ and take an inverse limit as $n \to \infty$ to get a homomorphism $\rho_{E,\ell} : G_L \to GL_2(\mathbb{Z}_\ell)$. By extending the scalars to $\mathbb{Q}_\ell$ or any of its extension fields by tensoring, one gets a 2-dimensional representation of G_L over an ℓ-adic field. This homomorphism has a number of important properties. First it is unramified almost everywhere. As the field fixed by the kernel of $\rho_{E,\ell}$ is an infinite extension of L, this last statement requires some interpretation. Let J be the field fixed by the kernel of $\rho_{E,\ell}$. The extension J/L is infinite, but for any finite M/L with $M \subset J$ the relative discriminant of $\text{Dis}_{M/K}$ is relatively prime to all but a fixed finite set of prime ideals of L depending only on E and ℓ. The primes of L outside this set are called *unramified*. Taking the inverse limit, this implies there is, for every unramified prime ideal $\mathfrak{p}$ in J/L, a conjugacy class $\text{Frob}_\mathfrak{p}$ associated to $\text{Gal}(J/L) \subset GL_2(\mathbb{Z}_\ell)$. Its characteristic polynomial, $\det(I - \text{Frob}_\mathfrak{p} X)$, is then well-defined. A priori this polynomial is in $\mathbb{Z}_\ell[X]$ but one can in fact prove it is in $\mathbb{Z}[X]$. It contains important information about E, namely that the coefficient of X, denoted $-a_\mathfrak{p}$, determines the number of points of the elliptic curve mod $\mathfrak{p}$, that is,

$$\#E(\mathbb{F}_\mathfrak{p}) = \#\mathbb{F}_\mathfrak{p} + 1 - a_\mathfrak{p}.$$

These polynomials, as we vary $\mathfrak{p}$, determine the local at $\mathfrak{p}$ factors of the L-function of E (except for the ramified primes where the local L-factor is slightly different), which according to the Birch and Swinnerton-Dyer Conjecture determines the rank of the abelian group $E(L)$, of L-rational points of E.

EXAMPLE 5.2. For the elliptic curve $E : y^2 = x^3 + 1$, it is known that for each prime $p > 3$,

$$(5.1) \qquad \#E(\mathbb{F}_p) = \begin{cases} p + 1 + J(\phi, \eta_3) + \overline{J(\phi, \eta_3)} & \text{if } p \equiv 1 \pmod{3} \\ p + 1 & \text{if } p \equiv 2 \pmod{3} \end{cases},$$

where ϕ is the unique quadratic character on $\mathbb{F}_p^\times$ and η_3 is a cubic character of $\mathbb{F}_p^\times$, see [**46**, §18.3]. Furthermore E has complex multiplication by $\mathbb{Z}[\zeta_3]$, that is, its endomorphim ring is an order in $\mathbb{Z}[\zeta_3]$ properly containing $\mathbb{Z}$. These 'extra' endomorphisms are generated over $\mathbb{Z}$ by $(x, y) \mapsto (\zeta_3 x, y)$. When viewing E as a variety over $\mathbb{Q}$, these endomorphisms are necessarily defined over $\mathbb{Q}(\zeta_3)$. The image of $G_{\mathbb{Q}(\zeta_3)}$ under $\rho_{E,\ell} : G_{\mathbb{Q}(\zeta_3)} \to GL_2(\mathbb{Z}_\ell)$ commutes with these extra endomorphisms. It is then easy to see that the image of $\rho_{E,\ell}|_{G_{\mathbb{Q}(\zeta_3)}}$ is abelian. When $\ell \equiv 1 \pmod{3}$ (the ordinary case) one finds that $\rho_{E,\ell}|_{G_{\mathbb{Q}(\zeta_3)}}$ is a direct sum of two characters. For $\ell \equiv 2 \pmod{3}$ one must extend the coefficients from $\mathbb{Z}_\ell$ to a larger ring to get the direct sum.

5.2. Grössencharacters in the sense of Hecke

We now recall a result of Weil which is relevant to our discussion below. Weil computed the local zeta functions for (homogeneous) Fermat curves of the form $X^n + Y^n = Z^n$ or special generalized Legendre curves of the form $y^N = x^{m_1}(1-x)^{m_2}$ (cyclic covers of $\mathbb{C}P^1$ only ramify at $0, 1, \infty$). In both cases the local zeta functions can be expressed in terms of explicit Jacobi sums, see [**100**] by Weil or the textbook [**46**] by Ireland and Rosen. In [**101**], Weil explained how to consider Jacobi sums as Grössencharacters (also written 'Grössencharakter' by Hecke and Weil and sometimes called Hecke characters) which we recall below. Here, we mainly use Weil's notation below.

DEFINITION 5.3. [See Weil [**101**]] Let L be a number field with r_1 non-equivalent real embeddings and r_2 non-equivalent complex embeddings. Fix an embedding of L to $\mathbb{C}$ and let $\mathcal{O}_L$ be its ring of integers as before. Let $\mathfrak{m}$ be a nonzero ideal of $\mathcal{O}_L$ and use $\mathcal{I}(\mathfrak{m})$ to denote the set of ideals of $\mathcal{O}_L$ that are prime to $\mathfrak{m}$. A *Grössencharacter* (or 'Grössencharakter') of L with defining ideal $\mathfrak{m}$, according to Hecke, is a complex-valued function f on the set of ideals of $\mathcal{O}_L$ such that
 (1) $f(\mathfrak{a})f(\mathfrak{b}) = f(\mathfrak{ab})$ for all $\mathfrak{a}, \mathfrak{b} \in \mathcal{I}(\mathfrak{m})$
 (2) There are rational integers e_i and rational numbers c_i, with $1 \leq i \leq r_1 + r_2$, such that if $a \in \mathcal{O}_L$, $a \equiv 1 \pmod{\mathfrak{m}}$ and a is positive at all real embeddings $L \hookrightarrow \mathbb{R}$, then the value of f at the principal ideal (a) satisfies

$$f((a)) = \prod_{i=1}^{r_1+r_2} a_i^{e_i} |a_i|^{c_i},$$

where $a_1 = a, a_2, \cdots, a_{r_1+r_2}$ are the non-equivalent embeddings of a to $\mathbb{C}$, $|\cdot|$ denotes the complex absolute value, and we use the principal branch for $|a_i|^{c_i}$.

The defining ideal $\mathfrak{m}$ is not unique. The greatest common divisor of all defining ideals is called the *conductor* of f and is also a defining ideal.

REMARK 5.4. In fact, the above two conditions being satisfied simultaneously imposes many restrictions on the e_i's and c_j's. For instance, the condition (1) implies that if a is a unit which is 1 modulo $\mathfrak{m}$, then $f((a)) = 1$. From this one deduces a dependence relation on the $|a_i|$ with the e_i's and c_j's as coefficients.

EXAMPLE 5.5. Let $L = \mathbb{Q}$ and $\mathfrak{m} = (1)$, then the set $\mathcal{I}(\mathfrak{m})$ contains all ideals of $\mathbb{Z}$, which are all principal. For $I \in \mathcal{I}(\mathfrak{m})$, write $I = (a)$ with $a > 0$. Define a function $\mathcal{T}$ on $\mathcal{I}(\mathfrak{m})$ by $\mathcal{T}(I) = a$. Then $\mathcal{T}$ is a Grössencharacter of $\mathbb{Q}$ with $r_1 = 1, r_2 = 0$, $e_1 = 0$ and $c_1 = 1$.

EXAMPLE 5.6. Let $L = \mathbb{Q}(\sqrt{-1})$ with $r_1 = 0, r_2 = 1$ and let $\mathfrak{m} = (4)$. The ring of integers of L is $\mathbb{Z}[i]$, a PID. Define a function f on prime ideals $(a+b\sqrt{-1}) \in \mathcal{I}(\mathfrak{m})$ as follows

$$f((a+b\sqrt{-1})) = (a+b\sqrt{-1})\chi(a+b\sqrt{-1}),$$

where

$$\chi(a+b\sqrt{-1}) = \begin{cases} (-1)^{\frac{b+1}{2}} \sqrt{-1}, & \text{if } a \equiv 0 \pmod{2}, b \equiv 1 \pmod{2}, \\ (-1)^{\frac{a-1}{2}}, & \text{if } a \equiv 1 \pmod{2}, b \equiv 0 \pmod{2}, \\ 0, & \text{otherwise.} \end{cases}$$

It is straightforward to check that f is a multiplicative function on ideals in $\mathcal{I}(\mathfrak{m})$. Also, when $a + b\sqrt{-1} \equiv 1 \pmod{\mathfrak{m}}$, $f((a+b\sqrt{-1})) = a + b\sqrt{-1}$. So f is a Grössenchacter of L with defining ideal (4). In this case, $e_1 = 1$ and $c_1 = 0$.

REMARK 5.7. For any fixed number field L, by definition, the set of Grössencharacters is closed under multiplication and division. Two Grössencharacters are equivalent if they agree on all ideals coprime to both defining ideals of the characters. The classes of equivalent Grössencharacters are in bijective correspondence with the homomorphisms from the group of idele classes of L to $\mathbb{C}^\times$. In general, these characters, according to Definition 5.3 need not take value in $S^1 = \{z \in \mathbb{C} : |z| = 1\}$. By class field theory, finite order characters of $G_L = \mathrm{Gal}(\overline{L}/L)$ correspond to finite order Grössencharacters of L. In this case, we sometime use the same notation for both the Grössencharacter and the corresponding character of G_L.

5.3. Notation for the Nth power residue symbol

Let N be a natural number and set $K = \mathbb{Q}(\zeta_N)$. Recall that $\mathcal{O}_K$ denotes the ring of integers of K. For each finite prime ideal $\mathfrak{p} \subset \mathcal{O}_K$ that is coprime to N, let $q(\mathfrak{p}) := \#(\mathcal{O}_K/\mathfrak{p})$ and let $\mathbb{F}_\mathfrak{p}$ be the residue field. Necessarily, $q(\mathfrak{p}) \equiv 1 \pmod{N}$. We define a map from $\mathcal{O}_K$ to the set of Nth roots of unity together with 0 using the following Nth power residue symbol notation $\left(\dfrac{x}{\mathfrak{p}}\right)_N$ (here we simply call it the Nth symbol), see pp. 240-241 of [**74**] by Milne.

DEFINITION 5.8. For $x \in \mathcal{O}_K$ define $\left(\dfrac{x}{\mathfrak{p}}\right)_N := 0$ if $x \in \mathfrak{p}$, and if $x \notin \mathfrak{p}$, let the symbol take the value of the unique Nth root of unity such that

$$(5.2) \qquad \left(\frac{x}{\mathfrak{p}}\right)_N \equiv x^{(q(\mathfrak{p})-1)/N} \pmod{\mathfrak{p}}.$$

We can extend the definition to $x \in K$, provided $\mathfrak{p}$ does not appear to a negative power in the factorization of the fractional ideal (x).

For explicit examples of the Nth symbols with $N = 2, 3, 4, 6$, see [**46**].

For a fixed unramified prime ideal $\mathfrak{p}$, the Nth symbol induces a multiplicative map from the residue field $\mathbb{F}_\mathfrak{p}$ to $\mathbb{C}$ which sends $0 \in \mathbb{F}_\mathfrak{p}$ to 0 and thus is compatible with our notion of multiplicative characters on $\mathbb{F}_\mathfrak{p}$ introduced in §2.2.

DEFINITION 5.9. Now for any fixed rational number of the form $\frac{i}{N}$ with $i, N \in \mathbb{Z}$, we define a map $\iota_{(\cdot)}\left(\frac{i}{N}\right)$ from the set of unramified prime ideals $\mathfrak{p}$ of $\mathcal{O}_K$ to multiplicative characters of the corresponding residue fields $\mathcal{O}_K/\mathfrak{p}$ by

$$(5.3) \qquad \iota_\mathfrak{p}\left(\frac{i}{N}\right)(\cdot) = \left(\frac{\cdot}{\mathfrak{p}}\right)_N^i.$$

By definition, for any integers i, j, N,

$$\iota_\mathfrak{p}\left(\frac{i+j}{N}\right) = \iota_\mathfrak{p}\left(\frac{i}{N}\right)\iota_\mathfrak{p}\left(\frac{j}{N}\right), \quad \text{and} \quad \overline{\iota_\mathfrak{p}\left(\frac{i}{N}\right)} = \iota_\mathfrak{p}\left(\frac{-i}{N}\right),$$

where the bar denotes complex conjugation as before.

EXAMPLE 5.10. For each unramified prime ideal with residue field of odd characteristic, $\iota_{\mathfrak{p}}(\frac{1}{2}) = \phi$, the quadratic character.

Fix a $c \in K^\times$ and let $\mathfrak{p}$ be a prime ideal of $\mathcal{O}_K$ such that $\mathfrak{p}$ is prime to N and $\operatorname{ord}_{\mathfrak{p}}(c) = 0$. Let $\operatorname{Frob}_{\mathfrak{p}}$ denote the Frobenius automorphism associated to $\mathfrak{p}$ in the abelian Kummer extension $K(c^{1/N})/K$. Using the Artin symbol one easily checks that for varying $\mathfrak{p}$

$$(5.4) \qquad \iota_{\mathfrak{p}}\left(\frac{1}{N}\right)(c) = \frac{\operatorname{Frob}_{\mathfrak{p}}(c^{1/N})}{c^{1/N}}.$$

For $c \in K^\times$, we extend the Nth symbol $\left(\frac{c}{\cdot}\right)_N$ multiplicatively to get a map from the set of ideals of $\mathcal{O}_K$ that are coprime to both c and $\operatorname{Dis}_{K/\mathbb{Q}}$ to the multiplicative group μ_N as follows. For each ideal $\mathfrak{a}$ of such, we decompose it as $\mathfrak{p}_1^{e_1} \cdots \mathfrak{p}_s^{e_s}$ where $\mathfrak{p}_i$ are distinct prime ideals. Then we let

$$\left(\frac{c}{\mathfrak{a}}\right)_N := \left(\frac{c}{\mathfrak{p}_1}\right)_N^{e_1} \cdots \left(\frac{c}{\mathfrak{p}_s}\right)_N^{e_s}.$$

See page 241 of [74]. When $\frac{i}{N} = \frac{1}{2}$, $K = \mathbb{Q}$, and $c \in \mathbb{Q}^\times$, this step means we extend the Legendre symbol $\left(\frac{c}{\cdot}\right) := \left(\frac{c}{\cdot}\right)_2$ to the Jacobi symbol.

PROPOSITION 5.11. *For $N \in \mathbb{N}$ and $c \in K^\times$ the map $\mathfrak{a} \mapsto \left(\frac{c}{\mathfrak{a}}\right)_N$ corresponds to a 1-dimensional representation of G_K and the kernel is $G_{K(\sqrt[N]{c})}$.*

PROOF. By the construction, the map $\mathfrak{a} \mapsto \left(\frac{c}{\mathfrak{a}}\right)_N$ is a Grössencharacter with finite image and we may take as a defining ideal the discriminant ideal $\operatorname{Dis}_{K(\sqrt[N]{c})/K}$. By Kummer theory, it corresponds to a 1-dimensional representation of G_K with kernel $G_{K(\sqrt[N]{c})}$. $\square$

If we fix $a = \frac{i}{N}$ and $c \in K^\times$, then for any prime ideal $\mathfrak{a}$ coprime to $\operatorname{Dis}_{K(\sqrt[N]{c})/K}$, the map

$$(5.5) \qquad \mathfrak{a} \mapsto \chi_{\frac{i}{N},c}(\mathfrak{a}) := \left(\frac{c}{\mathfrak{a}}\right)_N^i$$

also corresponds to a 1-dimensional representation of G_K with kernel being $G_{K(\sqrt[N]{c})}$. So the above map $\chi_{a,c}$ is a Grössencharacter, see Remark 5.7.

EXAMPLE 5.12. Let $K = \mathbb{Q}$, $c = -1$ in which case $r_1 = 0, r_2 = 1$, $N = 2$ and $\mathfrak{m} = (4)$. In this case, if $e_1 = 1$ and $c_1 = -1$ then condition (2) of Definition 5.3 holds for the map $\chi_{\frac{1}{2},-1}$.

Conversely, for any character $A \in \widehat{\mathbb{F}_{\mathfrak{p}}^\times}$ of order dividing N, we have another map $\kappa_{\mathbb{F}_{\mathfrak{p}}}$ which assigns a rational number to A. In order to define this map, we first recall that if $x \in \mathcal{O}_K \backslash \mathfrak{p}$ then $A(x) \in \mu_N$. We will show there exists $i \in \mathbb{N}$ such that $A(x) = \left(\frac{x}{\mathfrak{p}}\right)_N^i$ for all $x \in \mathcal{O}_K$. To see that fix an x_0 whose image generates the cyclic group $(\mathcal{O}_K/\mathfrak{p})^\times$. As the order of $A(x_0)$ divides N, which is the order of $\left(\frac{x_0}{\mathfrak{p}}\right)_N$, we have that $A(x_0) = \left(\frac{x_0}{\mathfrak{p}}\right)_N^i$ for some i. For any $x \in \mathcal{O}_K$ but not in $\mathfrak{p}$ we have, for some r, that $x \equiv x_0^r \pmod{\mathfrak{p}}$. Then for all x,

$$(5.6) \qquad A(x) = A(x_0^r) = A(x_0)^r = \left(\frac{x_0}{\mathfrak{p}}\right)_N^{ri} = \left(\frac{x_0^r}{\mathfrak{p}}\right)_N^i = \left(\frac{x}{\mathfrak{p}}\right)_N^i.$$

We may thus define $\kappa_{\mathbb{F}_\mathfrak{p}}(A) = \frac{i}{N}$. Then for any $a = \frac{i}{N}$, and any prime ideal $\mathfrak{p}$ of $\mathcal{O}_{\mathbb{Q}(\zeta_N)}$ coprime to N,
$$\kappa_{\mathbb{F}_\mathfrak{p}}(\iota_\mathfrak{p}(a)) \equiv a \pmod{\mathbb{Z}}.$$

Next we will see that the Nth symbol notation is compatible with field extensions. Recall that one can lift a multiplicative character $A \in \widehat{\mathbb{F}_q^\times}$ to any finite extension of $\mathbb{F}_q$ by using the norm map. Let L be a finite extension of $\mathbb{Q}(\zeta_N)$ and let $\wp$ be a prime ideal in the ring of integers $\mathcal{O}_L$ of L above $\mathfrak{p}$, with $\mathfrak{p}$ coprime to the discriminant of L. Then $\mathbb{F}_\wp := \mathcal{O}_L/\wp$ is a finite extension of $\mathbb{F}_\mathfrak{p} := \mathcal{O}_{\mathbb{Q}(\zeta_N)}/\mathfrak{p}$ and we denote the degree of the extension by f.

$$(5.7) \quad \left(\frac{\mathrm{N}_{\mathbb{F}_\mathfrak{p}}^{\mathbb{F}_\wp}(x)}{\mathfrak{p}}\right)_N = \left(\frac{x \cdot x^{q(\mathfrak{p})} \cdot x^{q(\mathfrak{p})^2} \cdots x^{q(\mathfrak{p})^{f-1}}}{\mathfrak{p}}\right)_N = \left(\frac{x^{\frac{q(\mathfrak{p})^f - 1}{q(\mathfrak{p})-1}}}{\mathfrak{p}}\right)_N$$
$$\equiv x^{(q(\mathfrak{p})^f - 1)/N} \pmod{\mathfrak{p}}.$$

But $x^{(q(\mathfrak{p})^f - 1)/N} = x^{(q(\wp)-1)/N} \equiv \left(\frac{x}{\wp}\right)_N \pmod{\wp}$. Thus on the residue field level, this means the $\left(\frac{\cdot}{\wp}\right)_N$ symbol on $\mathbb{F}_\wp$ can be computed from the norm map $\mathrm{N}_{\mathbb{F}_\mathfrak{p}}^{\mathbb{F}_\wp}$ composed with the map $\left(\frac{\cdot}{\mathfrak{p}}\right)_N$ on $\mathbb{F}_\mathfrak{p}$.

Using $\iota_\mathfrak{p}(\cdot)$ (see (5.3)), one can associate to any hypergeometric function
$$_{n+1}P_n\begin{bmatrix} a_1 & a_2 & \cdots & a_{n+1} \\ & b_1 & \cdots & b_n \end{bmatrix};\lambda\end{bmatrix}$$
such that $a_i, b_j, \lambda \in \mathbb{Q}$, a collection of hypergeometric functions over finite residue fields $\mathbb{F}_\mathfrak{p}$ (varying in $\mathfrak{p}$)
$$_{n+1}\mathbb{P}_n\begin{bmatrix} \iota_\mathfrak{p}(a_1) & \iota_\mathfrak{p}(a_2) & \cdots & \iota_\mathfrak{p}(a_{n+1}) \\ & \iota_\mathfrak{p}(b_1) & \cdots & \iota_\mathfrak{p}(b_n) \end{bmatrix};\lambda;q(\mathfrak{p})\end{bmatrix},$$
where $\mathfrak{p}$ runs through all unramified prime ideals of $\mathbb{Q}(\zeta_N)$ with N being the least positive common denominator of all a_i and b_j. We will see this explicitly for the $n=1$ case in §6.3.

5.4. Jacobi sums and Grössencharacters

Let $\mathfrak{p}$ be an unramified prime ideal of $K = \mathbb{Q}(\zeta_N)$. We will now give Weil's result (for his $r=2$ case). Let $\underline{a} = (\frac{a_1}{N}, \frac{a_2}{N})$ with $a_i \in \mathbb{Z}$. For each prime ideal $\mathfrak{p}$ coprime to N, let

$$\mathcal{J}_{\underline{a}}(\mathfrak{p}) := -\left(\frac{-1}{\mathfrak{p}}\right)_N^{a_1+a_2} \sum_{x \in \mathcal{O}_K/\mathfrak{p}} \left(\frac{x}{\mathfrak{p}}\right)_N^{a_1} \left(\frac{1-x}{\mathfrak{p}}\right)_N^{a_2}$$
$$= -\left(\frac{-1}{\mathfrak{p}}\right)_N^{a_1+a_2} J\left(\iota_\mathfrak{p}\left(\frac{a_1}{N}\right), \iota_\mathfrak{p}\left(\frac{a_2}{N}\right)\right).$$

Alternatively, one can write

$$(5.8) \qquad \mathcal{J}_{\underline{a}}(\mathfrak{p}) = -\chi_{\frac{a_1+a_2}{N},-1}(\mathfrak{p}) \cdot J\left(\iota_\mathfrak{p}\left(\frac{a_1}{N}\right), \iota_\mathfrak{p}\left(\frac{a_2}{N}\right)\right).$$

5.4. JACOBI SUMS AND GRÖSSENCHARACTERS

Next we extend $\mathcal{J}_{\underline{a}}$ to all ideals in $\mathcal{I}((N))$ by using
$$\mathcal{J}_{\underline{a}}(\mathfrak{ab}) = \mathcal{J}_{\underline{a}}(\mathfrak{a})\mathcal{J}_{\underline{a}}(\mathfrak{b}),$$
if $\mathfrak{a}, \mathfrak{b} \in \mathcal{I}((N))$.

THEOREM 5.13 (Weil, [**101**]). *The map $\mathcal{J}_{\underline{a}}$ is a Grössencharacter of $\mathbb{Q}(\zeta_N)$ with a defining ideal $\mathfrak{m} = (N^2)$.*

Namely, Weil showed that property (2) in Definition 5.3 above also holds for this map when $\mathfrak{m} = (N^2)$. Note that the conductor is not (N) in general. For example, when $\underline{a} = (\frac{1}{2}, 1)$, $\mathcal{J}_{\underline{a}}(\mathfrak{p}) = \chi_{\frac{1}{2}, -1}(\mathfrak{p})$ which was mentioned in Example 5.12. Its conductor is (4) instead of (2).

Sometimes quotients of Jacobi sums take values that are roots of unity.

EXAMPLE 5.14. By Example 2 of [**23**], we have that for $N = 10$ and $\mathfrak{p}$ any unramified prime ideal of $\mathcal{O}_{\mathbb{Q}(\zeta_{10})}$, the Grössencharacter $\mathcal{J}_{(\frac{1}{10}, \frac{6}{10})}/\mathcal{J}_{(\frac{2}{10}, \frac{5}{10})}(\mathfrak{p})$ of $\mathbb{Q}(\zeta_{10})$ satisfies
$$\mathcal{J}_{(\frac{1}{10}, \frac{6}{10})}/\mathcal{J}_{(\frac{2}{10}, \frac{5}{10})}(\mathfrak{p}) = \left(\frac{2}{\mathfrak{p}}\right)_{10}^8 = \chi_{\frac{8}{10}, 2}(\mathfrak{p}).$$

EXAMPLE 5.15. It is shown by Yamamoto (see [**104**, §20]) that the Grössencharacter $\mathcal{J}_{(\frac{2}{12}, \frac{5}{12})}/\mathcal{J}_{(\frac{3}{12}, \frac{4}{12})}(\mathfrak{p})$ of $\mathbb{Q}(\zeta_{12})$ is not a multiplicative character of $\mathbb{F}_{\mathfrak{p}}$ for all primes $\mathfrak{p}$ coprime to 12, but its square is. To be more precise, Yamamoto showed that
$$\left(\mathcal{J}_{(\frac{2}{12}, \frac{5}{12})}/\mathcal{J}_{(\frac{3}{12}, \frac{4}{12})}\right)^2(\mathfrak{p}) = \left(\frac{27/4}{\mathfrak{p}}\right)_{12} = \chi_{\frac{1}{12}, \frac{27}{4}}(\mathfrak{p}).$$

One reason for the above phenomenon, known as the sign ambiguity, is that in Weil's Theorem 5.13, a defining ideal is (N^2), but not (N) in general.

CHAPTER 6

Galois Representation Interpretation

In this chapter we use the theory discussed in Chapter 5 to interpret our finite field period function in terms of Galois representations, starting with the $_1\mathbb{P}_0$ functions in §6.1. Then we review the specific case associated to generalized Legendre curves in §6.2. In §6.3, we describe the Galois interpretation for the period functions $_2\mathbb{P}_1$ and prove Theorem 1.1. We then discuss the interpretation of the special cases of imprimitive $_2\mathbb{P}_1$ functions in §6.4, and the Galois interpretation of the normalized $_2\mathbb{F}_1$ functions in §6.5. In §6.6 we discuss how finite field period functions can give information about local zeta functions for hypergeometric varieties and give a few examples.

6.1. Galois interpretation for $_1\mathbb{P}_0$

To illustrate our ideas, we will first discuss the Galois representation background behind the $_1\mathbb{P}_0$ function defined in (4.3) using the $\iota_{(\cdot)}$ map (see (5.3)). Recall that by our notation given in (3.3),

$$_1P_0\left[\frac{i}{N};\lambda\right] = (1-\lambda)^{-\frac{i}{N}}.$$

Let $\lambda \in \mathbb{Q}$ and $K = \mathbb{Q}(\zeta_N)$. Now fix $\frac{i}{N}, \lambda \in \mathbb{Q}^\times$ and allow $\mathfrak{p}$ to vary. For each prime ideal $\mathfrak{p}$ of $\mathcal{O}_K$ coprime to the discriminant ideal of $K(\sqrt[N]{1-\lambda})/\mathbb{Q}$, we have

$$_1\mathbb{P}_0\left[\iota_\mathfrak{p}\left(\frac{i}{N}\right);\lambda;q(\mathfrak{p})\right] = \overline{\iota_\mathfrak{p}\left(\frac{i}{N}\right)(1-\lambda)} = \iota_\mathfrak{p}\left(-\frac{i}{N}\right)(1-\lambda) = \chi_{-\frac{i}{N},1-\lambda}(\mathfrak{p}),$$

by the definition in (4.3). On the other hand, if we fix $\mathfrak{p}$ and let i vary, then by Proposition 5.11, $_1\mathbb{P}_0[\iota_\mathfrak{p}(\cdot);\lambda;q(\mathfrak{p})]$ provides a map which sends rational numbers of the form $\frac{i}{N}$ to finite order characters of the Galois group $G_{\mathbb{Q}(\zeta_N)}$ whose kernel contains $G_{\mathbb{Q}(\sqrt[N]{1-\lambda},\zeta_N)}$. Further one can compute the corresponding Artin L-function

$$(6.1) \quad L\left(\frac{i}{N},\lambda;s\right)$$

$$= \prod_{\text{good } \mathfrak{p} \text{ of } \mathcal{O}_K} \left(1 - {}_1\mathbb{P}_0\left[\iota_\mathfrak{p}\left(\frac{i}{N}\right);\lambda;q(\mathfrak{p})\right]q(\mathfrak{p})^{-s}\right)^{-1}.$$

Note the right side only includes those prime ideals $\mathfrak{p}$ that are coprime to the absolute discriminant of $\mathbb{Q}(\sqrt[N]{1-\lambda},\zeta_N)$.

EXAMPLE 6.1. For $\frac{i}{N} = \frac{1}{2}$ and $\lambda = -1$ as in Example 5.12, we have

$$L\left(\frac{1}{2},-1;s\right) = \prod_{p \text{ odd prime}} \left(1 - \left(\frac{-1}{p}\right)p^{-s}\right)^{-1}.$$

6.2. Generalized Legendre curves and their Jacobians

We now describe the Galois interpretation of our finite field period functions for the specific setting corresponding to generalized Legendre curves and their Jacobians.

For any complex numbers a, b, c, z, with $\mathrm{Re}(c) > \mathrm{Re}(b) > 0$, we have that the formula of Euler (see [6])

$$\int_0^1 x^{b-1}(1-x)^{c-b-1}(1-zx)^{-a}dx$$

converges, and the classical hypergeometric series ${}_2F_1\begin{bmatrix} a & b \\ & c \end{bmatrix}; z\end{bmatrix}$ can be expressed, for $|z| < 1$, as

$${}_2F_1\begin{bmatrix} a & b \\ & c \end{bmatrix}; z\end{bmatrix} = \frac{1}{B(b, c-b)} \int_0^1 x^{b-1}(1-x)^{c-b-1}(1-zx)^{-a}dx,$$

where the branch is determined by

$$\arg(x) = 0,\ \arg(1-x) = 0,\ |\arg(1-zx)| < \frac{\pi}{2},\ x \in (0,1),$$

and $B(\cdot, \cdot)$ is the beta function defined in Definition 2.2. As described in §2.1, the restriction $\mathrm{Re}(c) > \mathrm{Re}(b) > 0$ needed for $B(\cdot, \cdot)$ can be dropped if we take the Pochhammer contour γ_{01} (see Definition 2.3 for the notation for γ_{ab}) as the integration path in the definition of $B(\cdot, \cdot)$.

If the parameters $a, b, c \in \mathbb{Q}$, and $a, b, a-c, b-c \notin \mathbb{Z}$, which correspond to the non-degenerate cases, Wolfart [103] realized that the integrals

$$\frac{1}{(1-e^{2\pi i b})(1-e^{2\pi i(c-b)})} \int_{\gamma_{01}} x^{b-1}(1-x)^{c-b-1}(1-\lambda x)^{-a}dx$$

$$= B(a,b)\, {}_2F_1\begin{bmatrix} a & b \\ & c \end{bmatrix}; \lambda\end{bmatrix} = {}_2P_1\begin{bmatrix} a & b \\ & c \end{bmatrix}; \lambda\end{bmatrix}$$

and

(6.2) $$\frac{1}{(1-e^{-2\pi i a})(1-e^{2\pi i(c-a)})} \int_{\gamma_{\frac{1}{\lambda}\infty}} x^{b-1}(1-x)^{c-b-1}(1-\lambda x)^{-a}dx$$

$$= (-1)^{c-a-b-1}\lambda^{1-c} B(1+a-c, 1-a)\, {}_2F_1\begin{bmatrix} 1+b-c & 1+a-c \\ & 2-c \end{bmatrix}; \lambda\end{bmatrix}$$

$$= (-1)^{c-a-b-1}\lambda^{1-c}\, {}_2P_1\begin{bmatrix} 1+b-c & 1+a-c \\ & 2-c \end{bmatrix}; \lambda\end{bmatrix}$$

are both *periods* (related to differential 1-forms of the form (6.7) below) of a so-called *generalized Legendre curve* of the form

(6.3) $$y^N = x^i(1-x)^j(1-\lambda x)^k,$$

where

(6.4) $$N = \mathrm{lcd}(a,b,c),\quad i = N \cdot (1-b),\quad j = N \cdot (1+b-c),\quad k = N \cdot a,$$

6.2. GENERALIZED LEGENDRE CURVES AND THEIR JACOBIANS

and lcd means the least (positive) common denominator. By the assumption that $a, b, a-c, b-c \notin \mathbb{Z}$, we know $N \nmid i, j, k, i+j+k$. This assumption is to require that the curve be a cover of $\mathbb{C}P^1$ ramifying at exactly the four distinct points $0, 1, 1/\lambda, \infty$ ($\lambda \neq 0, 1$). Also by changing variables if needed, one can assume $0 < i, j, k < N$.

EXAMPLE 6.2. In Example 3.5, we see that for the order 2 hypergeometric differential equation with parameters $a = 1/6, b = 1/3, c = 5/6$, the projective monodromy group is isomorphic to the arithmetic triangle group $(3, 6, 6)$. Using (6.4), one can compute that $N = 6, i = 4, j = 3, k = 1$ for this case. Similarly, the projective monodromy group for the hypergeometric differential equation with parameters $a = 1/12, b = 1/4, c = 5/6$ is isomorphic to the arithmetic triangle group $(2, 6, 6)$. For the latter case, by (6.4), $N = 12, i = 9, j = 5, k = 1$.

For $a = b = \frac{1}{2}, c = 1$ and any fixed $\lambda \in \mathbb{Q}$ with $\lambda \neq 0, 1$, the corresponding curve is the well-known Legendre curve

$$L_\lambda : y^2 = x(1-x)(1-\lambda x), \tag{6.5}$$

see [86]. We summarize a few relevant properties of the Legendre curves here.

- It is a double over of $\mathbb{C}P^1$ which ramifies only at $0, 1, \frac{1}{\lambda}, \infty$ as demonstrated by the picture below. Going from right to left, first cut the torus twice including half of each of the indicated boundaries on each torus, to get two cylinders. Each cylinder can be realized as the sphere on the left by pinching the ends together. Gluing along the slits gives the double cover.

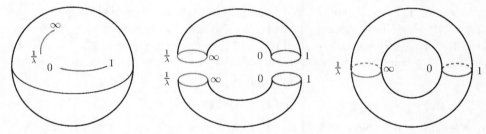

- As $\lambda \in \mathbb{Q} \setminus \{0, 1\}$, the curve L_λ is an algebraic curve defined over $\mathbb{Q}$ with genus 1.
- Every holomorphic differential on L_λ is a scalar multiple of

$$\omega_\lambda := \frac{dx}{\sqrt{x(1-x)(1-\lambda x)}}.$$

- A period of L_λ is $2 \cdot \int_0^1 \omega_\lambda = 2 \cdot {}_2P_1 \begin{bmatrix} \frac{1}{2} & \frac{1}{2} \\ & 1 \end{bmatrix}; \lambda = 2\pi \cdot {}_2F_1 \begin{bmatrix} \frac{1}{2} & \frac{1}{2} \\ & 1 \end{bmatrix}; \lambda$. As described in section §3.2.1, the hypergeometric function ${}_2F_1 \begin{bmatrix} \frac{1}{2} & \frac{1}{2} \\ & 1 \end{bmatrix}; \lambda$ satisfies the differential equation $HDE(\frac{1}{2}, \frac{1}{2}; 1; \lambda)$ and thus the monodromy group of $HDE(\frac{1}{2}, \frac{1}{2}; 1; \lambda)$ is isomorphic to the arithmetic triangle group (∞, ∞, ∞) as described in §3.2.4.
- For simplicity fix $\lambda \in \mathbb{Q} \setminus \{0, 1\}$. As we recalled in §5.1 (see also [85, 86]), there is a compatible family of 2-dimensional ℓ-adic representations $\{\rho_{L_\lambda, \ell}\}$ of $G_\mathbb{Q}$ constructed from the Tate module of L_λ. For any prime ℓ and for any prime $p \neq \ell$ not dividing the discriminant of the elliptic curve

L_λ (denoted by $N(L_\lambda)$), $\rho_{\lambda,\ell}$ is unramified at p. Thus it makes sense to consider the trace and determinant of $\rho_{\lambda,\ell}$ evaluated at the conjugacy class of Frobenius at p. In particular,

$$\operatorname{Tr}\rho_{L_\lambda,\ell}(\operatorname{Frob}_p) = -\sum_{x\in\mathbb{F}_p}\phi(x(1-x)(1-\lambda x)) = -{}_2\mathbb{P}_1\begin{bmatrix}\phi & \phi \\ & \varepsilon\end{bmatrix};\lambda\end{bmatrix}$$

and $\det\rho_{L_\lambda,\ell}(\operatorname{Frob}_p) = p$, where Frob_p stands for the Frobenius conjugacy class of p in $G_\mathbb{Q}$.

- For $\lambda \in \mathbb{Q}\setminus\{0,1\}$, the L-function of L_λ is

(6.6) $$L(L_\lambda, s) \text{``=''} \prod_{p\nmid N(L_\lambda)}\left(1 + {}_2\mathbb{P}_1\begin{bmatrix}\phi & \phi \\ & \varepsilon\end{bmatrix};\lambda\end{bmatrix}p^{-s} + p^{1-2s}\right)^{-1}.$$

Here, the quotations indicate that we are only giving a formula for the good L-factors.

For more general cases, Archinard in [8] explained how to construct the smooth model $X_\lambda^{[N;i,j,k]}$ of $C_\lambda^{[N;i,j,k]}$ for $\lambda \neq 0, 1$ of (6.3) and to compute all periods of first and second kind on $C_\lambda^{[N;i,j,k]}$ using hypergeometric functions. This is also recast in [23], and we follow that development here. Like L_λ, this curve is also a cyclic cover of $\mathbb{C}P^1$ ramifying at $0, 1, \frac{1}{\lambda}$, and ∞. When $\lambda = 0$ or 1, the number of ramification points of the covering map is at most 3 and hence the covering curve is a quotient of a Fermat curve. This is the reason behind the degenerate situation happening at $\lambda = 0$ or 1. Note that when $N \mid i, j, k$, or $i+j+k$, one can rewrite $C_\lambda^{[N;i,j,k]}$ as $y^N = x^i(1-x)^j$ by changing variables. We exclude this degenerate case below.

To consider the generic case of $C_\lambda^{[N;i,j,k]}$ being a cyclic cover of $\mathbb{C}P^1$ ramifying at exactly 4 points, we will assume the following for the remaining of this Chapter:

$$N \nmid i, j, k, i+j+k, \quad \gcd(i,j,k,N) = 1, \quad \lambda \neq 0, 1,$$

where gcd stands for greatest common divisor.

REMARK 6.3. Note that the assumptions $N \nmid i, j, k, i+j+k$ are satisfied if i, j, k, N are computed by (6.4) from any given $(a, b, c) \in \mathbb{Q}^3$ satisfying $a, b, c-a, c-b \notin \mathbb{Z}$. Also, by changing variables, one can assume $i, j, k > 0$.

Assume $\lambda \in \mathbb{Q}\setminus\{0,1\}$. Then $X_\lambda^{[N;i,j,k]}$ has genus

$$g(N;i,j,k) := 1 + N - \frac{\gcd(N,i+j+k) + \gcd(N,i) + \gcd(N,j) + \gcd(N,k)}{2},$$

see [8] by Archinard. We use $J_\lambda^{[N;i,j,k]}$ to denote the Jacobian of $X_\lambda^{[N;i,j,k]}$, which is an abelian variety of dimension given by $g(N;i,j,k)$ and is defined over $\mathbb{Q}$. For each proper divisor d of N, there is a curve $C_\lambda^{[d;i,j,k]}$ and a canonical map $C_\lambda^{[N;i,j,k]} \to C_\lambda^{[d;i,j,k]}$ sending (x,y) to $(x, y^{N/d})$ which is generically of degree N/d. This canonical map induces a surjective homomorphism $\pi_d : J_\lambda^{[N;i,j,k]} \to J_\lambda^{[d;i,j,k]}$. We now use $J_\lambda^{\mathrm{prim}}$ to denote the primitive part of $J_\lambda^{[N;i,j,k]}$, the identity component of $\bigcap_{d\mid N}\ker\pi_d$.

The curve $C_\lambda^{[N;i,j,k]}$ admits an automorphism $A_{\zeta_N} : (x,y) \mapsto (x, \zeta_N^{-1}y)$ and this map induces a representation of the finite group $\mathbb{Z}/N\mathbb{Z}$, depending on the choice of

6.2. GENERALIZED LEGENDRE CURVES AND THEIR JACOBIANS

the primitive Nth root ζ_N, on the vector space $H^0(X_\lambda^{[N;i,j,k]}, \Omega^1)$ of the holomorphic differential 1-forms on $X_\lambda^{[N;i,j,k]}$. When N, i, j, k, λ are fixed, we denote this curve by $X(\lambda)$ below for simplicity. A basis of $H^0(X(\lambda), \Omega^1)$ can be chosen by the regular pull-backs of differentials on $C_\lambda^{[N;i,j,k]}$ of the form

$$(6.7) \qquad \omega = \frac{x^{b_0}(1-x)^{b_1}(1-\lambda x)^{b_2} dx}{y^n}, \quad 0 \leq n \leq N-1, \, b_i \in \mathbb{Z},$$

satisfying the following conditions equivalent to the pullback of ω being regular at $0, 1, \frac{1}{\lambda}, \infty$ respectively,

$$b_0 \geq \frac{ni + \gcd(N,i)}{N} - 1, \quad b_1 \geq \frac{nj + \gcd(N,j)}{N} - 1, \quad b_2 \geq \frac{nk + \gcd(N,k)}{N} - 1,$$

$$b_0 + b_1 + b_2 \leq \frac{n(i+j+k) - \gcd(N, i+j+k)}{N} - 1.$$

For details on this construction, see the work of Archinard and Wolfart in [8, 103]. For each $0 \leq n < N$, we let V_n denote the isotypical component of $H^0(X(\lambda), \Omega^1)$ associated to the character $\sigma_n : \zeta_N \mapsto \zeta_N^n$, where ζ_N is a primitive Nth root of unity. Then the space $H^0(X(\lambda), \Omega^1)$ is decomposed into a direct sum $\bigoplus_{n=0}^{N-1} V_n$. If $\gcd(n, N) = 1$, the dimension of V_n is given by

$$\dim V_n = \left\{\frac{ni}{N}\right\} + \left\{\frac{nj}{N}\right\} + \left\{\frac{nk}{N}\right\} - \left\{\frac{n(i+j+k)}{N}\right\},$$

where $\{x\} = x - \lfloor x \rfloor$ denotes the fractional part of x, see [9]. Furthermore,

$$\dim V_n + \dim V_{N-n} = 2,$$

when $\gcd(n, N) = 1$. The elements of V_n with $\gcd(n, N) = 1$ are said to be *new*. The subspace

$$H^0(X(\lambda), \Omega^1)^{\text{new}} = \bigoplus_{\gcd(n,N)=1} V_n$$

is of dimension $\varphi(N)$, Euler's totient function of N, see [8].

Let S be a basis of $H^0(X(\lambda), \Omega^1)^{\text{new}}$ whose elements are of the form $\omega_n = x^{b_0}(1-x)^{b_1}(1-\lambda x)^{b_2} dx/y^n$ with $\gcd(N, n) = 1$. Under our assumptions, J_λ^{prim} is of dimension $\varphi(N)$, and is defined over $\mathbb{Q}$.

The Jacobian variety J_λ^{prim} is isomorphic to the quotient of $\mathbb{C}^{\varphi(N)}$ by the lattice of periods and it is isogenous to the complex torus $\mathbb{C}^{\varphi(N)}/\Lambda(\lambda)$ with

$$\Lambda(\lambda) = \left\{ \left(\sigma_n(u) \int_{\gamma_{01}} \omega + \sigma_n(v) \int_{\gamma_{\frac{1}{\lambda}\infty}} \omega \right)_{\omega \in S} : u, v \in \mathbb{Z}[\zeta_N] \right\},$$

by Archinard and Wolfart [8, 103]. Here $\gcd(n, N) = 1$, and σ_n is the automorphism of $\mathbb{Z}[\zeta_N]$ defined by $\zeta_N \mapsto \zeta_N^n$.

In the case of $0 < i, j, k < N$ and $N < i+j+k < 2N$, we have

$$\dim V_n = \dim V_{N-n} = 1, \quad \gcd(N, n) = 1,$$

and ω_n is $x^{-\{ni/N\}}(1-x)^{-\{nj/N\}}(1-\lambda x)^{-\{nk/N\}} dx$. For instance, the differential forms ω_1 and ω_{N-1} are $\omega_1 = dx/y$ and $\omega_{N-1} = \frac{x^{i-1}(1-x)^{j-1}(1-\lambda x)^{k-1}}{y^{N-1}} dx$. Thus, when $1 \leq i, j, k < N$, $\gcd(N, i, j, k) = 1$, $N \nmid i+j$ nor $i+j+k$, and $\lambda \neq 0, 1$, the lattice $\Lambda(\lambda)$ can be expressed in terms of

$$_2P_1 \left[\begin{array}{cc} \{\frac{nk}{N}\} & 1 - \{\frac{ni}{N}\} \\ & 2 - \{\frac{ni}{N}\} - \{\frac{nj}{N}\} \end{array} ; \lambda \right].$$

However, when $0 < i + j + k < N$ or $2N < i + j + k < 3N$, we do not have such a general form for the vector space V_n.

EXAMPLE 6.4. The spaces V_1 and V_3 corresponding to the family $C_\lambda^{[4;1,1,1]}$ have dimension 0 and 2, respectively. The space $H^0\left(X(\lambda), \Omega^1\right)$ is spanned by

$$S = \{dx/y^2, \, dx/y^3, \, xdx/y^3\},$$

and hence

$$_2P_1 \left[\begin{array}{cc} \frac{1}{2} & \frac{1}{2} \\ & 1 \end{array} ; \lambda \right], \, _2P_1 \left[\begin{array}{cc} \frac{3}{4} & \frac{1}{4} \\ & \frac{1}{2} \end{array} ; \lambda \right], \, _2P_1 \left[\begin{array}{cc} \frac{3}{4} & \frac{5}{4} \\ & \frac{3}{2} \end{array} ; \lambda \right]$$

are periods of $C_\lambda^{[4;1,1,1]}$.

For the family $C_\lambda^{[5;3,4,4]}$, we have $\dim V_1 = 2$ and $\dim V_2 = 1$. A basis of $H^0\left(X(\lambda), \Omega^1\right)$ is

$$S = \left\{ \frac{dx}{y}, \, \frac{xdx}{y}, \, \frac{x(1-x)(1-\lambda x)dx}{y^2}, \, \frac{x(1-x)^2(1-\lambda x)^2 dx}{y^3} \right\}.$$

6.3. Galois interpretation for $_2\mathbb{P}_1$

In this section we prove Theorem 1.1. Let $\lambda \in \mathbb{Q} \setminus \{0, 1\}$ and set $K = \mathbb{Q}(\zeta_N)$ and $\overline{K}$ to be its algebraic closure.

For any fixed prime ℓ, similar to the elliptic curve case discussed in §5.1, the ℓ^n-torsion points of the abelian variety J_λ^{prim} gives rise to a continuous homomorphism $\rho_{\lambda,\ell}^{\text{prim}}$ from group $\text{Gal}(\overline{K}/K)$ to $GL_{2\varphi(N)}(\mathbb{Z}_\ell)$. For simplicity, we extend the scalar rings to $\overline{\mathbb{Q}}_\ell$ and note that $\rho_{\lambda,\ell}^{\text{prim}}$ only ramifies at finite many places. Recall that we write η_N to denote any character of order N on $\mathbb{F}_q^\times$ and extend η_N to be defined on $\mathbb{F}_q$ by setting $\eta_N(0) = 0$. Evaluating the traces of the representation $\rho_{\lambda,\ell}^{\text{prim}}$ at $\text{Frob}_\mathfrak{p}$ for any unramified prime $\mathfrak{p}$ of $\mathcal{O}_K$ with $q(\mathfrak{p}) = q$, one has

$$(6.8) \quad \text{Tr}\, \rho_{\lambda,\ell}^{\text{prim}}(\text{Frob}_\mathfrak{p}) = - \sum_{m \in (\mathbb{Z}/N\mathbb{Z})^\times} \left(\sum_{x \in \mathbb{F}_q} \eta_N^m \left(x^i(1-x)^j(1-\lambda x)^k \right) \right)$$

$$= - \sum_{m \in (\mathbb{Z}/N\mathbb{Z})^\times} {}_2\mathbb{P}_1 \left[\begin{array}{cc} \eta_N^{-mk} & \eta_N^{mi} \\ & \eta_N^{m(i+j)} \end{array} ; \lambda; q \right].$$

For a related discussion, see Proposition 4.2; for more details, see [**23**]. Essentially, the equation (6.8) can be proved by using induction on N.

We are now ready to prove Theorem 1.1.

PROOF OF THEOREM 1.1. Fix an embedding of ζ_N to $\overline{\mathbb{Q}}_\ell$. Observe that the map A_{ζ_N}, which is defined over K, induces an order N automorphism $A_{\zeta_N}^*$ on

$J_\lambda^{\mathrm{prim}}$ and hence the representation spaces of $\rho_{\lambda,\ell}^{\mathrm{prim}}$ over $\overline{\mathbb{Q}}_\ell$ by our assumption. Consequently, by the construction of $J_\lambda^{\mathrm{prim}}$

$$\rho_{\lambda,\ell}^{\mathrm{prim}}|_{G_K} \cong \bigoplus_{\gcd(m,N)=1} \sigma_{\lambda,m},$$

where $\sigma_{\lambda,m}$ corresponds to the ζ_N^m eigenspace of $A_{\zeta_N}^*$. Due to the symmetric roles of $\sigma_{\lambda,m}$, they have the same dimension, which has to be 2.

For any $c \in K^\times$, fix $\sqrt[N]{c}$ an Nth root. We consider the smooth model of the following twisted generalized Legendre curve

$$C_{\lambda,c}^{[N;i,j,k]}: \quad y^N = cx^i(1-x)^j(1-\lambda x)^k.$$

It is isomorphic to $C_\lambda^{[N;i,j,k]}$ via the map $T_c : C_{\lambda,c}^{[N;i,j,k]} \to C_\lambda^{[N;i,j,k]}$ defined by $(x,y) \mapsto (x, \sqrt[N]{c}\,y)$. Note that the primitive part of its Jacobian variety $J_{\lambda,c}^{\mathrm{prim}}$ is also $\varphi(N)$ dimensional defined over $\mathbb{Q}$. Let $\rho_{\lambda,c,\ell}^{\mathrm{prim}}$ denote the corresponding Galois representation of G_K over $\overline{\mathbb{Q}}_\ell$, and note its restriction over G_K also decomposes into a direct sum of 2-dimensional subrepresentations, denoted by $\sigma_{\lambda,c,m}$ like the case $c=1$ before. To proceed, we fix a large prime ℓ and consider the ℓ^n division points on the Jacobian $J_\lambda^{\mathrm{prim}}$. Identify the Jacobian variety J_λ of the curve $C_\lambda^{[N;i,j,k]}$ with the group $Pic^0(C_\lambda^{[N;i,j,k]})$ of the classes of divisors of degree zero on $C_\lambda^{[N;i,j,k]}$. Assume that P represents the class $\sum_{i=1}^s n_i(x_i, y_i)$ where $(x_i, y_i) \in C_{\lambda,c}^{[N;i,j,k]}$ and degree $\sum_{i=1}^s n_i = 0$ and P is an ℓ^n-division point on $J_{\lambda,c}^{\mathrm{prim}}$. We further assume that as an element in the group algebra $\overline{\mathbb{Q}}_\ell[J_{\lambda,c}[\ell^n]]$, P lies in the ζ_N^m-eigenspace of the automorphism on $\rho_{\lambda,c,\ell}^{\mathrm{prim}}[\ell^n]$ induced from $A_{\zeta_N} : (x,y) \mapsto (x, \zeta_N^{-1} y)$ where $(m,N)=1$. The isomorphism T_c sends ℓ^n-division points on $J_{\lambda,c}^{\mathrm{prim}}$ to ℓ^n-division points on $J_\lambda^{\mathrm{prim}}$. For any $\mathrm{Frob}_\mathfrak{p} \in G_K$ with $\mathfrak{p}$ coprime to the discriminant of $K(\sqrt[N]{c})$,

$$\begin{aligned}
\mathrm{Frob}_\mathfrak{p}(T_c(P)) &= \mathrm{Frob}_\mathfrak{p} \sum_{i=1}^s n_i\left(x_i, \sqrt[N]{c}\cdot y_i\right) \\
&\stackrel{(5.4)}{=} \sum_{i=1}^s n_i\left(\mathrm{Frob}_\mathfrak{p}(x_i), \left(\frac{c}{\mathfrak{p}}\right)_N \sqrt[N]{c}\cdot \mathrm{Frob}_\mathfrak{p}(y_i)\right) \\
&= T_c\left(\sum_{i=1}^s n_i\left(\mathrm{Frob}_\mathfrak{p}(x_i), \left(\frac{c}{\mathfrak{p}}\right)_N \cdot \mathrm{Frob}_\mathfrak{p}(y_i)\right)\right) \\
&= T_c\left(\frac{c}{\mathfrak{p}}\right)_N^{-m} P.
\end{aligned}$$

This means $\sigma_{\lambda,c,m} \cong \sigma_{\lambda,m} \otimes \chi_{-\frac{m}{N},c}$ where $\chi_{-\frac{m}{N},c}$ is as in (5.5). Summing up all pieces, we have

$$\mathrm{Tr}\rho_{\lambda,c,\ell}^{\mathrm{prim}}(\mathrm{Frob}_\mathfrak{p}) = \sum_{m \in (\mathbb{Z}/N\mathbb{Z})^\times} \left(\frac{c}{\mathfrak{p}}\right)_N^{-m} \cdot \mathrm{Tr}\sigma_{\lambda,m}(\mathrm{Frob}_\mathfrak{p}).$$

Meanwhile, in terms of explicit point counting using characters,

$$\mathrm{Tr}\rho_{\lambda,c,\ell}^{\mathrm{prim}}(\mathrm{Frob}_{\mathfrak{p}}) = -\sum_{m\in(\mathbb{Z}/N\mathbb{Z})^\times}\sum_{x\in\mathbb{F}_{\mathfrak{p}}}\iota_{\mathfrak{p}}\left(\frac{m}{N}\right)(cx^i(1-x)^j(1-\lambda x)^k)$$

$$= -\sum_{m\in(\mathbb{Z}/N\mathbb{Z})^\times}\left(\frac{c}{\mathfrak{p}}\right)_N^m \cdot \sum_{x\in\mathbb{F}_{\mathfrak{p}}}\iota_{\mathfrak{p}}\left(\frac{m}{N}\right)(x^i(1-x)^j(1-\lambda x)^k).$$

As the above two equations hold for arbitrary $c \in K$, we have

$$\mathrm{Tr}\sigma_{\lambda,m}(\mathrm{Frob}_{\mathfrak{p}}) = -\sum_{x\in\mathbb{F}_{\mathfrak{p}}}\iota_{\mathfrak{p}}\left(\frac{-m}{N}\right)(x^i(1-x)^j(1-\lambda x)^k)$$

$$= -{}_2\mathbb{P}_1\left[\begin{matrix}\iota_{\mathfrak{p}}(\frac{mk}{N}) & \iota_{\mathfrak{p}}(\frac{-mi}{N}) \\ & \iota_{\mathfrak{p}}(\frac{-m(i+j)}{N})\end{matrix};\lambda\right].$$

Using (6.4), if we let $\sigma_{\lambda,1}$ above be the 2-dimensional representation $\sigma_{\lambda,\ell}$ stated in Theorem 1.1, then $\iota_{\mathfrak{p}}(\frac{mk}{N}) = \iota_{\mathfrak{p}}(a)$, $\iota_{\mathfrak{p}}(\frac{-mi}{N}) = \iota_{\mathfrak{p}}(b)$, and $\iota_{\mathfrak{p}}(\frac{-m(i+j)}{N}) = \iota_{\mathfrak{p}}(c)$ respectively, which concludes the proof of Theorem 1.1. □

REMARK 6.5. In other words, for fixed rational numbers a, b, c, the corresponding

$$-{}_2\mathbb{P}_1\left[\begin{matrix}\iota_{\mathfrak{p}}(a) & \iota_{\mathfrak{p}}(b) \\ & \iota_{\mathfrak{p}}(c)\end{matrix};\lambda;q(\mathfrak{p})\right]$$

functions are the traces (or characters) of an explicit 2-dimensional Galois representation $\sigma_{\lambda,\ell}$ of G_K at the Frobenius elements. When $a, b, c-a, c-b \notin \mathbb{Z}$, the representation is *pure* in the sense that for each good prime ideal $\mathfrak{p}$ such that $\lambda \in \mathbb{Z}$ and $\lambda \neq 0, 1 \pmod{p}$, the characteristic polynomial of $\sigma_{\lambda,\ell}(\mathrm{Frob}_{\mathfrak{p}})$ is of the form

$$(6.9) \qquad H_{\mathfrak{p}}(T) = T^2 - \mathrm{Tr}\sigma_{\lambda,\ell}(\mathrm{Frob}_{\mathfrak{p}})T + \det\sigma_{\lambda,\ell}(\mathrm{Frob}_{\mathfrak{p}})$$

and has two roots of the same absolute value $\sqrt{q}$ where $q = \#(\mathcal{O}_K/\mathfrak{p})$. See Corollary 6.11 for how to compute $\det\sigma_{\lambda,\ell}(\mathrm{Frob}_{\mathfrak{p}})$. When either $a, b, c-a$, or $c-b$ in $\mathbb{Z}$, the corresponding representation degenerates. We will give some examples in §6.4.

EXAMPLE 6.6. The following analogue of Kummer's evaluation (see (3.14)) expresses the value of the ${}_2\mathbb{P}_1$ function at -1 in terms of not one but two Jacobi sums. To be more precise, let $B, D, \phi \in \widehat{\mathbb{F}_q^\times}$ where ϕ is of order 2. Then, for $C = D^2$,

$$(6.10) \qquad {}_2\mathbb{P}_1\left[\begin{matrix}B & C \\ & C\overline{B}\end{matrix};-1;q\right] = J(D,\overline{B}) + J(D\phi,\overline{B}).$$

This is proved by Greene in [41, (4.11)]. For given $b, c \in \mathbb{Q}$ such that $b, c, c-2b \notin \mathbb{Z}$ and letting M be the least positive common denominator of $\frac{c}{2}, \frac{c+1}{2}, b$, there is a 2-dimensional ℓ-adic Galois representation $\sigma_{-1,\ell}$ of $G_{\mathbb{Q}(\zeta_N)}$ corresponding to ${}_2F_1\left[\begin{matrix}b & c \\ & c-b\end{matrix};-1\right]$ via Theorem 1.1 such that at each good unramified prime $\mathfrak{p}$,

$$\mathrm{Tr}\sigma_{-1,\ell}(\mathrm{Frob}_{\mathfrak{p}}) = \chi_{\frac{c}{2}-b,-1}(\mathfrak{p})\mathcal{J}_{(\frac{c}{2},-b)}(\mathfrak{p}) + \chi_{\frac{c+1}{2}-b,-1}(\mathfrak{p})\mathcal{J}_{(\frac{c+1}{2},-b)}(\mathfrak{p}),$$

where the notation $\mathcal{J}_{(a,b)}(\mathfrak{p}) = -\iota_{\mathfrak{p}}(a)\iota_{\mathfrak{p}}(b)(-1)J(\iota_{\mathfrak{p}}(a),\iota_{\mathfrak{p}}(b))$ as (5.8) in §5.4.

When $\lambda = 0$ or 1, the $_2\mathbb{P}_1$ function corresponds to a dimension 1 (instead of 2) compatible family of Galois representations and in these cases the corresponding generalized Legendre curves have smaller genus. When $\lambda = 1$, the curve $C_\lambda^{[N;i,j,k]}$ becomes $y^N = x^i(1-x)^{j+k}$ which is a quotient of a Fermat curve. The following analogue of Gauss' evaluation formula (3.13) follows from (4.5)

$$(6.11) \qquad {}_2\mathbb{P}_1 \begin{bmatrix} A & B \\ & C \end{bmatrix}; 1; q \end{bmatrix} = \sum_{y \in \mathbb{F}_q} B(y)\overline{AB}C(1-y) = J(B, \overline{AB}C).$$

They allow one to determine the Galois representation up to semisimplification and compute the determinants of the representations at Frobenius elements, see (6.15) below. Consequently, one can use these period functions to compute the Euler p-factors of the L-function of the corresponding Galois representations, as in equations (6.1) and (6.6). Watkins has written a Magma program which computes these p-factors under additional assumptions (namely when the hypergeometric motives are defined over $\mathbb{Q}$, see [**99**]. Also, in [**80**], there are discussions on the properties of such p-factors, hypergeometric L-functions, their conductors and functional equations.)

REMARK 6.7. From the representation point of view, the products (resp. sums) of $_2\mathbb{P}_1$ functions over the same finite field correspond to tensor products (resp. direct sums) of the corresponding representations. See [**84**].

6.4. Some special cases of $_2\mathbb{P}_1$-functions

In Remark 6.5, it is mentioned that when $a, b, a-c, b-c \notin \mathbb{Z}$, i.e. when $_2F_1 \begin{bmatrix} a & b \\ & c \end{bmatrix}; \lambda \end{bmatrix}$ is primitive (see Definition 4.3), the representation $\sigma_{\lambda,\ell}$ stated in Theorem 1.1 is pure. We now consider the cases below giving identities for the imprimitive cases, which correspond to $_2F_1 \begin{bmatrix} a & b \\ & c \end{bmatrix}; \lambda \end{bmatrix}$ with either $a, b, a-c, b-c \in \mathbb{Z}$. Indeed, in the first two formulas below without δ terms, the right sides can be thought of as 'q plus a unit' and thus resemble Eisenstein series corresponding to reducible representations. The last two formulas are also of this form when $\lambda \neq 1$ and resemble characters when $\lambda = 1$.

PROPOSITION 6.8. *Suppose* $\lambda \neq 0$. *Then*,

$${}_2\mathbb{P}_1 \begin{bmatrix} \varepsilon & B \\ & C \end{bmatrix}; \lambda \end{bmatrix} = J(B, \overline{B}C) - \overline{C}(\lambda)\overline{B}C(\lambda - 1);$$

$${}_2\mathbb{P}_1 \begin{bmatrix} A & B \\ & B \end{bmatrix}; \lambda \end{bmatrix} = \overline{B}(\lambda)J(B, \overline{A}) - \overline{A}(1-\lambda);$$

$${}_2\mathbb{P}_1 \begin{bmatrix} A & B \\ & A \end{bmatrix}; \lambda \end{bmatrix} = \overline{B}(\lambda - 1)J(B, \overline{A}) - B(-1)\overline{A}(\lambda) + (q-1)\delta(1-\lambda)\delta(B);$$

$${}_2\mathbb{P}_1 \begin{bmatrix} A & \varepsilon \\ & C \end{bmatrix}; \lambda \end{bmatrix} = \overline{C}(-\lambda)\overline{A}C(1-\lambda)J(C, \overline{A}) - 1 + (q-1)\delta(1-\lambda)\delta(\overline{A}C).$$

PROOF. When $\lambda \neq 0$, by (4.5)

$$_2\mathbb{P}_1\begin{bmatrix} \varepsilon & B \\ & C \end{bmatrix}; \lambda\end{bmatrix} = \sum_y B(y)\overline{B}C(1-y)\varepsilon(1-\lambda y) = \sum_{y \neq 1/\lambda} B(y)\overline{B}C(1-y)$$

$$= \sum_y B(y)\overline{B}C(1-y) - \overline{C}(\lambda)\overline{B}C(\lambda-1) = J(B, \overline{B}C) - \overline{C}(\lambda)\overline{B}C(\lambda-1);$$

$$_2\mathbb{P}_1\begin{bmatrix} A & B \\ & B \end{bmatrix}; \lambda\end{bmatrix} = \sum_y B(y)\varepsilon(1-y)\overline{A}(1-\lambda y) = \sum_{y \neq 1} B(y)\overline{A}(1-\lambda y)$$

$$= \overline{B}(\lambda)J(B, \overline{A}) - \overline{A}(1-\lambda).$$

Taking $A_1 = B_1 = A$ and $A_2 = B$ in (4.6) we have the function $_2\mathbb{P}_1\begin{bmatrix} A & B \\ & A \end{bmatrix}; \lambda\end{bmatrix}$ can be written as

$$_2\mathbb{P}_1\begin{bmatrix} A & B \\ & A \end{bmatrix}; \lambda\end{bmatrix} = \frac{B(-1)}{q-1}\sum_\chi J(A\chi, \overline{\chi})J(B\chi, \overline{A\chi})\chi(\lambda).$$

The relation
$$J(R, \overline{S})J(T, \overline{R}) = S(-1)J(T, \overline{S})J(T\overline{S}, \overline{R}S) - \delta(R)(q-1) + \delta(S\overline{T})(q-1)$$
can be verified using (2.10) through (2.15) and routine algebra. It is simplest to deal with the cases where the various δ terms are nonzero first.

Now take $R = A\chi$, $S = \chi$ and $T = B\chi$ to see our $_2\mathbb{P}_1$ is

$$\frac{B(-1)}{q-1}J(B, \overline{A})\left[\left(\sum_\chi J(B\chi, \overline{\chi})\chi(-\lambda)\right)\right] - B(-1)\overline{A}(\lambda) + \delta(B)\delta(1-\lambda)(q-1)$$

which, upon replacing $J(B\chi, \overline{\chi})$ by its defining sum and interchanging the order of summation in the resulting double sum, becomes

$$B(-1)J(B, \overline{A})\overline{B}(1-\lambda) - B(-1)\overline{A}(\lambda) + \delta(B)\delta(1-\lambda)(q-1)$$
$$= J(B, \overline{A})\overline{B}(\lambda-1) - B(-1)\overline{A}(\lambda) + \delta(B)\delta(1-\lambda)(q-1).$$

Using the same type of argument, we get the last claim. □

6.5. Galois interpretation for $_{n+1}\mathbb{F}_n$

We now use Theorem 1.1 to give a description of the Galois representations associated to the period functions $_2\mathbb{P}_1$. Recall $K = \mathbb{Q}(\zeta_N)$.

As mentioned earlier, when the parameter $\lambda \in \mathbb{Q}$, we have that the period function $_{n+1}\mathbb{P}_n$ corresponds to the trace of a Galois representation of degree at most n. The normalized period, which is obtained by dividing $_{n+1}\mathbb{P}_n$ by a product of Jacobi sums, corresponds to the tensor product of the Galois representation associated with $_{n+1}\mathbb{P}_n$ and a linear character associated with the Grössencharacter arising from the Jacobi sums as in §5.4.

When $n = 1$ and $\lambda \in \mathbb{Q} \setminus \{0, 1\}$, by Theorem 1.1 we have under the same assumptions, that there is a 2-dimensional ℓ-adic Galois representation

$$\tilde{\sigma}_{\lambda,\ell} := \sigma_{\lambda,\ell} \otimes (\chi_{c,-1} \cdot \mathcal{J}_{(b,c-b)})^{-1}$$

over $\overline{\mathbb{Q}}_\ell$ such that for any good prime ideal $\mathfrak{p}$ of $\mathcal{O}_K$,

$$\text{Tr}\,\tilde{\sigma}_{\lambda,\ell}(\text{Frob}_\mathfrak{p}) = {}_2\mathbb{F}_1\left[\begin{matrix} \iota_\mathfrak{p}(a) & \iota_\mathfrak{p}(b) \\ & \iota_\mathfrak{p}(c) \end{matrix}; \lambda; q(\mathfrak{p})\right].$$

See the proof of Theorem 1.1 for the construction of $\sigma_{\lambda,\ell}$, and (5.5) (resp. (5.8)) for the notation for $\chi_{c,-1}$ (resp. $\mathcal{J}_{b,c-b}$). Note that the $-$ sign is not needed for the above formula due to the relation between $\mathcal{J}_{(\cdot,\cdot)}$ and $J(\cdot,\cdot)$ given in (5.8). Moreover, when $c-b, b, c \notin \mathbb{Z}$, the characteristic polynomial of $\tilde{\sigma}_{\lambda,\ell}(\text{Frob}_\mathfrak{p})$ is degree 2 with two roots of absolute value 1. This will be particularly convenient when taking tensor products.

6.6. Zeta functions and hypergeometric functions over finite fields

In this section, we relate our finite field period functions ${}_{n+1}\mathbb{P}_n$ to the local zeta functions of hypergeometric varieties, and demonstrate how Theorem 1.1 can be used to compute the L-function of associated Galois representations.

Given any hypersurface H_f defined by an algebraic affine (or projective) equation $f(x_1,\cdots,x_n) = 0$ over a finite field $\mathbb{F}_q$, use N_s to denote the number of affine (projective) solutions of f over $\mathbb{F}_{q^s}$. The *zeta function* of H_f over $\mathbb{F}_q$ is defined by

$$Z_q(H_f, T) := \exp\left(\sum_{s=1}^\infty \frac{N_s T^s}{s}\right).$$

Now we recall the following theorem of Dwork.

THEOREM 6.9 (Dwork, see [50]). *The zeta function of any affine or projective hypersurface given by $f(x_1,\cdots,x_n) = 0$ is a ratio of two polynomials with coefficients in $\mathbb{Z}$ and constant 1.*

For example, if C is a smooth projective irreducible genus g curve defined over $\mathbb{F}_q$, then Weil showed that

$$Z_q(C,T) = \frac{P(T)}{(1-T)(1-qT)}$$

where $P(T) \in \mathbb{Z}[T]$ has degree $2g$ with all roots of absolute value $1/\sqrt{q}$. See [46] for more details.

In this perspective, Proposition 4.2 implies that the finite field period functions ${}_{n+1}\mathbb{P}_n$ are related to the local zeta functions of hypergeometric varieties given by algebraic equations of the form (1.1). In the following discussion, we focus on the case $n=1$. Along this line, we relate Theorem 1.1 to some result concerning twisted exponential sums. These involve the finite field period functions ${}_2\mathbb{P}_1$ whose parameters are characters on finite extensions of $\mathbb{F}_q$. Given $A, B, C \in \widehat{\mathbb{F}_q^\times}$, by our notation

$$ {}_2\mathbb{P}_1\left[\begin{matrix} A & B \\ & C \end{matrix}; \lambda; q\right] = \sum_{y \in \mathbb{F}_q} B(y)\overline{B}C(1-y)\overline{A}(1-\lambda y).$$

As mentioned in §5.3 any character A on $\mathbb{F}_q^\times$ can be extended to a multiplicative character A_r on the finite extension $\mathbb{F}_{q^r}$ using the norm map $\text{N}_{\mathbb{F}_q}^{\mathbb{F}_{q^r}}$, i.e. for $x \in \mathbb{F}_{q^r}$,

$$A_r(x) = A(\text{N}_{\mathbb{F}_q}^{\mathbb{F}_{q^r}}(x)).$$

See (5.7). For instance, if $x \in \mathbb{F}_q$, then

(6.12) $$A_r(x) = A(x \cdot x^q \cdots x^{q^{r-1}}) = A(x \cdot x \cdots x) = A(x^r) = (A(x))^r.$$

Thus we can define ${}_2\mathbb{P}_1 \left[\begin{smallmatrix} A_r & B_r \\ & C_r \end{smallmatrix} ; \lambda; q^r \right]$ accordingly. Using the perspective of twisted Jacobi sums [1,2], the generating function

(6.13) $$Z[A, B; C; \lambda; q; T] := \exp\left(\sum_{r \geq 1} \left({}_2\mathbb{P}_1 \left[\begin{smallmatrix} A_r & B_r \\ & C_r \end{smallmatrix} ; \lambda; q^r \right] \right) \cdot \frac{T^r}{r} \right)$$

is also a rational function, which is originally due to Dwork.

In view of Theorem 1.1 and the third Weil conjecture (proved by Deligne) for local zeta functions of smooth curves (cf. [46, §11.3]), we know for $a, b, c \in \mathbb{Q}$ and $a, b, c - a, c - b \notin \mathbb{Z}$, the corresponding Galois representation $\sigma_{\lambda,\ell}$ is pure and at each unramified prime ideal $\mathfrak{p}$ and $\lambda \neq 0, 1 \pmod{\mathfrak{p}}$, the function

$$Z[\iota_{\mathfrak{p}}(a), \iota_{\mathfrak{p}}(b); \iota_{\mathfrak{p}}(c); \lambda; q(\mathfrak{p}); T] = H_{\mathfrak{p}}\left(\frac{1}{T}\right)$$

where $H_{\mathfrak{p}}(T)$ is the characteristic polynomial of the Frobenius $\mathrm{Frob}_{\mathfrak{p}}$ under $\sigma_{\lambda,\ell}$, as in (6.9). Thus it is a polynomial with two roots of the same absolute value $1/\sqrt{q}$.

To give an example, we first recall another Hasse-Davenport relation (see Section 11.5 in [14] or Section 11.4 of [46]) which relates the Gauss sums $g(A)$ over $\mathbb{F}_q$ and $g(A_r)$ over $\mathbb{F}_{q^r}$ by

(6.14) $$g(A_r) = (-1)^{r-1} g(A)^r.$$

EXAMPLE 6.10. By (6.14) and the finite field analogue of Kummer's evaluation (6.10), one has when $C = D^2$, and $B^2 \neq C$, that

$$Z(B, C; C\overline{B}; -1; q; T) = (1 + J(D, \overline{B})T)(1 + J(D\phi, \overline{B})T).$$

In comparison, the finite field analogue of the Gauss evaluation formula (6.11) implies

$$Z[A, B; C; 1; q; T] = 1 + J(B, \overline{ABC})T, \quad \text{if } A \neq C.$$

As a corollary of Theorem 1.1, one can describe the determinants of the 2-dimensional Galois representation $\sigma_{\lambda,\ell}$ of Theorem 1.1 at Frobenius elements explicitly by a simple computation of eigenvalues of Frobenius elements.

COROLLARY 6.11. *Let the notation be as in Theorem 1.1. Then*
(6.15)
$$det\, \sigma_{\lambda,\ell}(Frob_{\mathfrak{p}}) = \frac{1}{2}\left(\left({}_2\mathbb{P}_1 \left[\begin{smallmatrix} \iota_{\mathfrak{p}}(a) & \iota_{\mathfrak{p}}(b) \\ & \iota_{\mathfrak{p}}(c) \end{smallmatrix}; \lambda; q(\mathfrak{p}) \right] \right)^2 + {}_2\mathbb{P}_1 \left[\begin{smallmatrix} (\iota_{\mathfrak{p}}(a))_2 & (\iota_{\mathfrak{p}}(b))_2 \\ & (\iota_{\mathfrak{p}}(c))_2 \end{smallmatrix}; \lambda; q(\mathfrak{p})^2 \right] \right),$$

utilizing the notation described in (6.12). *Together, we are able to compute the L-function* $L(\sigma_{\lambda,\ell}, s)$ *of* $\sigma_{\lambda,\ell}$ *if we omit the ramified prime factors. It will take the form of*

$$\prod_{\mathfrak{p} \text{ good prime}} \left(1 + {}_2\mathbb{P}_1 \left[\begin{smallmatrix} \iota_{\mathfrak{p}}(a) & \iota_{\mathfrak{p}}(b) \\ & \iota_{\mathfrak{p}}(c) \end{smallmatrix}; \lambda; q(\mathfrak{p}) \right] q(\mathfrak{p})^{-s} + det\, \sigma_{\lambda,\ell}(Frob_{\mathfrak{p}}) \cdot q(\mathfrak{p})^{-2s} \right)^{-1}.$$

6.7. Summary

To us, the Galois representation interpretation, i.e. that the $_{n+1}\mathbb{P}_n$ or $_{n+1}\mathbb{F}_n$ functions are, up to sign, the character values of Galois representations at corresponding Frobenius conjugacy classes, provides a useful guideline when one translates classical results to the finite field setting. A few principles are summarized as follows.

First, let $\lambda \in \mathbb{Q}$, and N the least common multiple of the orders of the A_i and B_j. Assume $q \equiv 1 \pmod{N}$ and that λ can be embedded into $\mathbb{F}_q$. Then:

- The function $_{n+1}\mathbb{F}_n \begin{bmatrix} A_1 & A_2 & \cdots & A_{n+1} \\ & B_1 & \cdots & B_n \end{bmatrix}; \lambda; q$ corresponds, up to a sign, to the trace of a degree at most $n+1$ Galois representation of G_K at the Frobenius conjugacy class at a prime ideal with residue of size q. When $\lambda = 0, 1$, the imprimitve cases, the representation is often degenerate.

- When $\lambda \neq 0, 1$, the primitive $_2\mathbb{P}_1$ or $_2\mathbb{F}_1$ function corresponds to a 2-dimensional Galois representation, which is pure.

- Products (resp. sums) of $_{n+1}\mathbb{P}_n$ or $_{n+1}\mathbb{F}_n$ functions correspond to tensor products (resp. direct sums) of the corresponding representations.

REMARK 6.12. In relation to the first item above, see [1, 2, 47] and the paper on hypergeometric motives by Roberts and Rodriguez-Villegas [80], as well as a different approach to realize the hypergeometric motives defined over $\mathbb{Q}$ by Beukers, Cohen, and Mellit in [16] based on Katz's result in [47]. See [16, 69] for some explicit examples in this approach.

CHAPTER 7

A finite field Clausen formula and an application

In this chapter, we use the previous discussion of Galois representations from Chapters 5 and 6 to discuss a finite field version of the classical Clausen formula, due to Evans and Greene, and how its geometric interpretation sheds light on Ramanujan type formulas for $1/\pi$. Below we continue to use ε to denote the trivial character and ϕ to denote the quadratic character. The discussion below is closely related to Weil's result describing Jacobi sums as Grössencharacters, which was recalled in §5.2.

7.1. A finite field version of the Clausen formula by Evans and Greene

One version of the Clausen formula [11] states that

$$(7.1) \qquad {}_2F_1\left[\begin{matrix} c-s-\frac{1}{2} & s \\ & c \end{matrix}; \lambda\right]^2 = {}_3F_2\left[\begin{matrix} 2c-2s-1 & 2s & c-\frac{1}{2} \\ & 2c-1 & c \end{matrix}; \lambda\right].$$

In terms of differential equations, this means that the symmetric square of the 2-dimensional solution space of $HDE(c-s-\frac{1}{2}, s; c; \lambda)$ (see (3.7)) is the 3-dimensional solution space of the hypergeometric differential equation satisfied by the ${}_3F_2$ occurring on the right side of (7.1). In [26], Evans and Greene obtained the following analogue of (7.1) written here in our notation.

THEOREM 7.1 ([26], Theorem 1.5). *Let* $C, S \in \widehat{\mathbb{F}_q^\times}$. *Assume that* $C \neq \phi$, *and* $S^2 \notin \{\varepsilon, C, C^2\}$. *Then for* $\lambda \neq 1$,

$${}_2\mathbb{F}_1\left[\begin{matrix} C\overline{S}\phi & S \\ & C \end{matrix}; \lambda\right]^2 = {}_3\mathbb{F}_2\left[\begin{matrix} C^2\overline{S}^2 & S^2 & C\phi \\ & C^2 & C \end{matrix}; \lambda\right]$$

$$+ \phi(1-\lambda)\overline{C}(\lambda)\left(\frac{J(\overline{S}^2, C^2)}{J(\overline{C}, \phi)} + \delta(C)(q-1)\right).$$

This theorem captures the well-known fact from representation theory that the tensor square of a 2-dimensional representation (associated with the ${}_2\mathbb{F}_1$ on the left) equals its symmetric square (3-dimensional representation, associated with the ${}_3\mathbb{F}_2$ on the right) plus an additional linear representation from the exterior power. The expression here using our notation is closer to the complex setting than the version of the statement given in Theorem 1.5 of [26] which is in terms of period functions.

We remark that when $\lambda = 1$, by Theorem 4.38-(i) in [**41**] we have

$$(7.2) \quad {}_3\mathbb{F}_2\left[\begin{matrix} C^2\overline{S}^2 & S^2 & C\phi \\ & C^2 & C \end{matrix}; 1\right] = \sum_{D\in\{S,S\phi\}} \frac{\phi(-1)J(D,C\overline{S}^2)J(\overline{C}S^2,\phi\overline{D})}{J(\phi,\phi C)J(S^2,C^2\overline{S}^2)}$$

$$= \sum_{D\in\{S,S\phi\}} \frac{J(D,C\overline{S}^2)J(\overline{C}S^2,\phi\overline{D})}{J(\phi S,\overline{C})J(S,\overline{C})},$$

which corresponds to a 2-dimensional Galois representation that can be described by Grössencharacters, while Gauss' evaluation theorem says that

$$ {}_2\mathbb{F}_1\left[\begin{matrix} C\overline{S}\phi & S \\ & C \end{matrix}; 1\right]^2 = \frac{J(\phi,S)^2}{J(S,\overline{C})^2}.$$

A nice example which realizes the Clausen formula geometrically is given by Ahlgren, Ono and Penniston [**5**]. In their work, they consider the $K3$ surfaces defined by

$$X_\lambda : s^2 = xy(1+x)(1+y)(x+\lambda y), \quad \lambda \neq 0, -1.$$

In particular, the point counting on X_λ over $\mathbb{F}_q$ is related to a ${}_3\mathbb{P}_2$ by the following equation,

$$(7.3) \quad \sum_{x,y\in\mathbb{F}_q} \phi(xy(1+x)(1+y)(x+\lambda y)) = \left[\begin{matrix} \phi & \phi & \phi \\ & \varepsilon & \varepsilon \end{matrix}; -\lambda\right].$$

One establishes the above equality by expanding the right hand side via (4.4) and employing the change of variables $(x,y) \mapsto (-x,-y)$ followed by $x \mapsto 1/x$.

The point counting on the following elliptic curve

$$E_\lambda : y^2 = (x-1)(x^2 - 1/(1+\lambda)), \quad \lambda \neq 0, -1,$$

over $\mathbb{F}_q$ is given by

$$a(\lambda, q) := -\sum_{x\in\mathbb{F}_q} \phi(x-1)\phi(x^2 - 1/(1+\lambda)).$$

Then we have the equality ([**5**, Theorem 2.1])

$$(7.4) \quad {}_3\mathbb{P}_2\left[\begin{matrix} \phi & \phi & \phi \\ & \varepsilon & \varepsilon \end{matrix}; -\lambda\right] = \phi(1+\lambda)(a(\lambda,q)^2 - q).$$

A geometric interpretation of the relation (7.4) between the K3 surfaces X_λ and the elliptic curve E_λ in terms of the so-called Shioda-Inose structure has been given in [**65**] by Long. When $q \equiv 1 \pmod{4}$, $a(\lambda, q)$ is essentially ${}_2\mathbb{P}_1\left[\begin{matrix} \eta_4 & \eta_4 \\ & \varepsilon \end{matrix}; -\lambda\right]$ where η_4 is an order 4 character. To be more precise, if $1+\lambda = b^2$ is a square in the finite field $\mathbb{F}_q$, then

$$a(\lambda, q) = \phi(b-1){}_2\mathbb{P}_1\left[\begin{matrix} \eta_4 & \eta_4 \\ & \varepsilon \end{matrix}; -\lambda\right].$$

Thus Theorem 1.1 of [**5**] is equivalent to the Clausen formula over the finite field $\mathbb{F}_q$ with $S = \eta_4$ and $C = \varepsilon$.

For special choices of $\lambda \in \mathbb{Q}$ such as $1, 8, 1/8, -4, -1/4$, the corresponding elliptic curve E_λ has complex multiplication (CM). For these λ values, the period functions ${}_2\mathbb{P}_1$ can be written in terms of Jacobi sums.

EXAMPLE 7.2. When $\lambda = 1/8$, the elliptic curve $E_{1/8}$ is $\mathbb{Q}$-isogenous to the curve $y^2 = x^3 - 9x$ which has CM and is of conductor 288 [**63**]. From this information and the explicit description of Grössencharacter $\mathcal{J}_{(\frac{1}{2},\frac{1}{4})}$, which is given in Example 5.6, one has

$$-a(1/8, q) = \phi(3) \cdot \begin{cases} 0, & \text{if } q \equiv -1 \pmod 4, \\ J(\phi, \eta_4) + J(\phi, \overline{\eta}_4), & \text{if } q \equiv 1 \pmod 4. \end{cases}$$

EXAMPLE 7.3. When $\lambda = -1/4$, the elliptic curve $E_{-1/4}$ is $\mathbb{Q}$-isogenous to the curve $y^2 = x^3 - 1$, which has complex multiplication and conductor 144 [**62**]. It follows that

$$-a(-1/4, q) = \phi(-1) \cdot \begin{cases} 0, & \text{if } q \equiv -1 \pmod 3 \\ J(\phi, \eta_3) + J(\phi, \overline{\eta}_3), & \text{if } q \equiv 1 \pmod 3, \end{cases}$$

where $-\phi(-1)J(\phi, \eta_3)$ is a Hecke character with conductor $\mathfrak{m} = (4\sqrt{-3})$ over $\mathbb{Q}(\zeta_6)$ and it can be given as follows. Suppose that a prime ideal $\mathfrak{p}$ of $\mathbb{Z}[\zeta_6]$ is generated by $\alpha \in \mathbb{Z}[\zeta_6]$. Then we have

$$\mathcal{J}_{(\frac{1}{2},\frac{1}{3})}(\mathfrak{p}) = \chi(\alpha) \cdot \alpha,$$

where $\chi(-1) = -1$, $\chi(5) = -1$, and $\chi(\zeta_6) = -\zeta_6$. Here, the unit group $(\mathbb{Z}[\zeta_6]/\mathfrak{m})^*$ has order 24 and is generated by -1, 5, and ζ_6.

7.2. Analogues of Ramanujan type formulas for $1/\pi$

In 1914, Ramanujan [**79**] gave a list of infinite series formulas for $1/\pi$, where the series are values of hypergeometric functions. One example is

(7.5) $$\sum_{k=0}^{\infty}(6k+1)\left(\frac{(\frac{1}{2})_k}{k!}\right)^3 \left(\frac{1}{4}\right)^k = \frac{4}{\pi}.$$

Later in the 1980's, Borwein-Borwein [**19**] and Chudnovsky-Chudnovsky [**21**] proved these formulas using essentially the theory of elliptic curves with complex multiplication. The idea of their proof was recast in [**20**] and we will give some related discussion here. Roughly speaking, the family of elliptic curves E_λ defined in §7.1 has unique holomorphic differentials ω_λ, up to scalars, similar to the case for the Legendre curve L_λ that we mentioned earlier in §6.2, see (6.5). Integrating ω_λ along a suitably chosen path on E_λ leads to a period of the first kind on E_λ given by

$$p(\lambda) = \Gamma\left(\frac{1}{4}\right)\Gamma\left(\frac{3}{4}\right)(1-\lambda)^{1/4} \,_2F_1\left[\begin{matrix}\frac{1}{4} & \frac{1}{4} \\ & 1\end{matrix}; -\lambda\right].$$

We know $\Gamma(\frac{1}{4})\Gamma(\frac{3}{4}) = \sqrt{2}\pi$ by the reflection formula, Theorem 2.4. One can compute periods of the second kind on E_λ from $\frac{d}{d\lambda}p(\lambda)$. Together, differentials of the first and second kind for E_λ form a 2-dimensional vector space $V(\lambda)$. When E_λ has CM, its endomorphism ring $R := R(\lambda)$, which induces an action on the space $V(\lambda)$, is larger than $\mathbb{Z}$. In particular, ω_λ is an eigenfunction of R. Let η_λ be a differential of the second kind that is also an eigenfunction of R. By Chowla-Selberg, for any cycle γ_λ of E_λ,

$$\int_{\gamma_\lambda} \omega_\lambda \cdot \int_{\gamma_\lambda} \eta_\lambda \sim \pi,$$

where $\sim$ means equality up to a multiple in $\overline{\mathbb{Q}}$. Picking γ_λ in the right way, one obtains a Ramanujan type formula for $1/\pi$ corresponding to the CM value λ. In this way, we obtain Ramanujan type formulas for $1/\pi$ using products of two distinct periods that are determined by R. In the spirit of the Clausen formula, the product of two periods lies in the symmetric square of the period space for E_λ; more explicitly, it is a special element in the solution space of the $_3F_2$.

While periods of the second kind obtained from derivation do not have exact finite field or Galois analogues, there are analogues of 'eigenfunctions' of the endomorphism ring on the Galois side. For suitable λ the elliptic curve E_λ has CM by a field we denote K_λ. Fix a prime ℓ and let $\rho_{\lambda,\ell}$ denote the family of 2-dimensional ℓ-adic Galois representations of $\mathrm{Gal}(\overline{\mathbb{Q}}/\mathbb{Q})$ constructed from the elliptic curve E_λ tensoring with $\mathbb{Q}_\ell$. As E_λ has CM,

$$\rho_{\lambda,\ell}|_{\mathrm{Gal}(\overline{\mathbb{Q}}/K_\lambda)} \simeq \sigma_{1,\lambda} \oplus \sigma_{2,\lambda},$$

where $\sigma_{1,\lambda}$ and $\sigma_{2,\lambda}$ are 1-dimensional representations that are invariant under R. Each of them corresponds to a Grössencharacter of K_λ, as we described explicitly for two cases above (Examples 7.2 and 7.3). The product $\sigma_{1,\lambda}\sigma_{2,\lambda}$ lies in the symmetric square of $\rho_{\lambda,\ell}|_{\mathrm{Gal}(\overline{\mathbb{Q}}/K_\lambda)}$. The Clausen formula implies it is also a linear factor of the 3-dimensional Galois representation corresponding to $_3\mathbb{P}_2$. For the special cases of λ (say $\lambda = -\frac{1}{4}, \frac{1}{8}$), one can describe the values of the $_3\mathbb{P}_2$ functions explicitly.

Using the above computation for $a(\lambda, q)$ and the Clausen formula, one has the following proposition based on Examples 7.2 and 7.3.

PROPOSITION 7.4. *We have the following two period function evaluation formulas.*

1. *For any prime power q that is coprime to 6, we have*

$$_3\mathbb{P}_2\begin{bmatrix} \phi & \phi & \phi \\ & \varepsilon & \varepsilon \end{bmatrix}; 1/4 \end{bmatrix} = \phi(-1)\begin{cases} q, & \text{if } q \equiv -1 \pmod{3}, \\ J(\phi, \eta_3)^2 + J(\phi, \overline{\eta}_3)^2 + q, & \text{if } q \equiv 1 \pmod{3}, \end{cases}$$

where η_3 is a character of order 3.

2. *For any prime power q that is coprime to 4, we have*

$$_3\mathbb{P}_2\begin{bmatrix} \phi & \phi & \phi \\ & \varepsilon & \varepsilon \end{bmatrix}; -1/8 \end{bmatrix} = \phi(-2)\begin{cases} q, & \text{if } q \equiv -1 \pmod{4}, \\ J(\phi, \eta_4)^2 + J(\phi, \overline{\eta}_4)^2 + q, & \text{if } q \equiv 1 \pmod{4}, \end{cases}$$

where η_4 is a character of order 4.

Here we remark that the values $\phi(-1)$ and $\phi(-2)$ respectively appearing on the right hand sides of the above equalities can be seen from (7.4) and the corresponding CM structures. See also Theorem 1 of [20]. Proposition 7.4 yields the following corollary.

COROLLARY 7.5. *When $\lambda = -1/4$ (resp. $\lambda = 1/8$), $\sigma_{1,\lambda}\sigma_{2,\lambda}$ extends to a 1-dimensional representation of $G_\mathbb{Q}$. It corresponds to the following Grössencharacter of $\mathbb{Q}$:*

$$\chi_{\frac{1}{2},-1} \cdot \mathcal{T} \quad (\text{resp. } \chi_{\frac{1}{2},-2} \cdot \mathcal{T}),$$

where $\mathcal{T}$ is defined in Example 5.5 and $\chi_{\frac{1}{2},(\cdot)}$ is defined in (5.5).

7.2. ANALOGUES OF RAMANUJAN TYPE FORMULAS FOR $1/\pi$

This is compatible with the p-adic analogues of Ramanujan-type formulas for $1/\pi$. For instance, the following supercongruence related to (7.5) was conjectured by van Hamme [95] and proved by the second author in [66]. For any prime $p > 3$,

$$\sum_{k=0}^{\frac{p-1}{2}}(6k+1)\left(\frac{(\frac{1}{2})_k}{k!}\right)^3\left(\frac{1}{4}\right)^k \equiv \left(\frac{-1}{p}\right) \pmod{p^4}.$$

See [20] for a general result, and [68] by the second and third authors, [90] by the fourth author, [77] by Osburn and Zudilin for some recent developments on Ramanujan-type supercongruences.

CHAPTER 8

Translation of Some Classical Results

In this chapter, we use our primary method to translate several classical results to their finite field analogues, while also making note of information gleaned from the corresponding Galois interpretations as appropriate. The results we translate include Kummer's 24 relations and the well-known Pfaff-Saalschütz formula, which were considered previously by Greene [**41**]. We additionally use our method to obtain a few finite field analogues of algebraic hypergeometric identities.

8.1. Kummer's 24 Relations

In this section we address the relations between the period $_2\mathbb{P}_1$ hypergeometric functions corresponding to independent solutions of the hypergeometric differential equation, and Kummer's 24 relations. Statements below without proof or further references are due to Greene, especially Theorem 4.4 in [**41**]. See also [**43**] by Greene for an interpretation in terms of representations of $\mathrm{SL}(2, q)$.

The following proposition corresponds to the discussion in §3.2.3 for the classical setting.

PROPOSITION 8.1. *For any characters $A, B, C \in \widehat{\mathbb{F}_q^\times}$, and $\lambda \in \mathbb{F}_q$, we have*

$$\begin{aligned}
{}_2\mathbb{P}_1\begin{bmatrix} A & B \\ & C \end{bmatrix};\lambda\end{bmatrix} &= ABC(-1)\overline{C}(\lambda)\,{}_2\mathbb{P}_1\begin{bmatrix} \overline{C}B & \overline{C}A \\ & \overline{C} \end{bmatrix};\lambda\end{bmatrix} + \delta(\lambda)J(B, C\overline{B}), \\
&= ABC(-1)\overline{A}(\lambda)\,{}_2\mathbb{P}_1\begin{bmatrix} A & \overline{C}A \\ & \overline{B}A \end{bmatrix};1/\lambda\end{bmatrix} + \delta(\lambda)J(B, C\overline{B}), \\
&= B(-1)\,{}_2\mathbb{P}_1\begin{bmatrix} A & B \\ & AB\overline{C} \end{bmatrix};1-\lambda\end{bmatrix}.
\end{aligned}$$

REMARK 8.2. In the first two equalities, the extra delta term on the right is necessary as when $\lambda = 0$ the left hand side is a non-zero Jacobi sum by definition, while the first term on the right hand side has value 0 by our convention. In the last entry, by replacing characters A, B, and C with characters AD, BD, and $\overline{C}D$, repectively, one gets

$${}_2\mathbb{P}_1\begin{bmatrix} AD & BD \\ & \overline{C}D \end{bmatrix};1-\lambda\end{bmatrix} = BD(-1)\,{}_2\mathbb{P}_1\begin{bmatrix} AD & BD \\ & ABCD \end{bmatrix};\lambda\end{bmatrix}.$$

Taking $\lambda = 0$ and applying (4.7) to both sides we obtain

$$\frac{BD(-1)}{q-1}\sum_\chi J(AD\chi, \overline{\chi})J(BD\chi, C\overline{D\chi}) = BD(-1)J(BD, AC).$$

By a variable change $\chi \mapsto \overline{D}\chi$ and the identity between the Gauss sums and Jacobi sums (2.13), we can deduce the identity of Helversen-Pasotto [45], which says that for any multiplicative characters A, B, C, D of $\mathbb{F}_q$

$$(8.1) \quad \frac{1}{q-1} \sum_{\chi} g(A\chi)g(B\chi)g(C\overline{\chi})g(D\overline{\chi}) =$$

$$\frac{g(AC)g(AD)g(BC)g(BD)}{g(ABCD)} + q(q-1)AB(-1)\delta(ABCD).$$

A representation theoretic proof of this formula is given in [60] by Li and Soto-Andrade. Using the representation theoretic approach, Li further obtained the p-adic field version of this identity in [58].

REMARK 8.3. Let $\lambda \in \mathbb{Q} \setminus \{0, 1\}$, and $a, b, c \in \mathbb{Q}$ such that $a, b, a-c, b-c, a+b-c \notin \mathbb{Z}$. Use $\sigma_{\lambda,1}$ (resp. $\sigma_{1-\lambda,2}$) to denote a 2-dimensional ℓ-adic Galois representation corresponding to ${}_2P_1\begin{bmatrix} a & b \\ & c \end{bmatrix}; \lambda$ (resp. ${}_2P_1\begin{bmatrix} a & b \\ & a+b-c \end{bmatrix}; 1-\lambda$) via Theorem 1.1. The last equality in Proposition 8.1 implies that if we use the $\chi_{(\cdot),(\cdot)}$ notation in (5.5), then

$${}_2\mathbb{P}_1\begin{bmatrix} \iota_{\mathfrak{p}}(a) & \iota_{\mathfrak{p}}(b) \\ & \iota_{\mathfrak{p}}(c) \end{bmatrix}; \lambda; q(\mathfrak{p}) = \chi_{b,-1}(\mathfrak{p}) \cdot {}_2\mathbb{P}_1\begin{bmatrix} \iota_{\mathfrak{p}}(a) & \iota_{\mathfrak{p}}(b) \\ & \iota_{\mathfrak{p}}(a+b-c) \end{bmatrix}; 1-\lambda; q(\mathfrak{p}).$$

This means $\sigma_{\lambda,1}$ is isomorphic to $\chi_{b,-1} \otimes \sigma_{1-\lambda,2}$ up to semisimplification. When $\lambda = 0$ or 1, both representations $\sigma_{\lambda,1}$ and $\sigma_{1-\lambda,2}$ have degree 1 instead of 2. Thus no extra delta terms will be needed in this case. Many other formulas in this section have similar Galois interpretations and most of them require additional delta terms.

The next proposition corresponds to the Pfaff and Euler transformations in §3.2.3.

PROPOSITION 8.4. For any characters $A, B, C \in \widehat{\mathbb{F}_q^\times}$, and $\lambda \in \mathbb{F}_q$, we have

$${}_2\mathbb{P}_1\begin{bmatrix} A & B \\ & C \end{bmatrix}; \lambda = \overline{A}(1-\lambda) \, {}_2\mathbb{P}_1\begin{bmatrix} A & \overline{B}C \\ & C \end{bmatrix}; \frac{\lambda}{\lambda-1} + \delta(1-\lambda)J(B, C\overline{AB}),$$

$$= \overline{B}(1-\lambda) \, {}_2\mathbb{P}_1\begin{bmatrix} C\overline{A} & B \\ & C \end{bmatrix}; \frac{\lambda}{\lambda-1} + \delta(1-\lambda)J(B, C\overline{AB}),$$

$$= \overline{ABC}(1-\lambda) \, {}_2\mathbb{P}_1\begin{bmatrix} \overline{A}C & \overline{B}C \\ & C \end{bmatrix}; \lambda + \delta(1-\lambda)J(B, C\overline{AB}).$$

The following relations between the normalized ${}_2\mathbb{F}_1$-hypergeometric functions follow immediately from Propositions 8.1 and 8.4.

8.1. KUMMER'S 24 RELATIONS

PROPOSITION 8.5. *For any characters* $A, B, C \in \widehat{\mathbb{F}_q^\times}$, *and* $\lambda \in \mathbb{F}_q$, *we have*

$$\begin{aligned}
{}_2\mathbb{F}_1\begin{bmatrix} A & B \\ & C \end{bmatrix}; \lambda\end{bmatrix} &= ABC(-1)\overline{C}(\lambda)\frac{J(\overline{C}A, \overline{A})}{J(B, C\overline{B})}\,{}_2\mathbb{F}_1\begin{bmatrix} \overline{C}B & \overline{C}A \\ & \overline{C} \end{bmatrix}; \lambda\end{bmatrix} + \delta(\lambda), \\
&= ABC(-1)\overline{A}(\lambda)\frac{J(\overline{C}A, \overline{B}C)}{J(B, C\overline{B})}\,{}_2\mathbb{F}_1\begin{bmatrix} A & \overline{C}A \\ & \overline{B}A \end{bmatrix}; 1/\lambda\end{bmatrix} + \delta(\lambda), \\
&= \frac{J(B, C\overline{AB})}{J(B, C\overline{B})}\,{}_2\mathbb{F}_1\begin{bmatrix} A & B \\ & AB\overline{C} \end{bmatrix}; 1-\lambda\end{bmatrix}, \\
&= \overline{A}(1-\lambda)\,{}_2\mathbb{F}_1\begin{bmatrix} A & \overline{B}C \\ & C \end{bmatrix}; \frac{\lambda}{\lambda-1}\end{bmatrix} + \delta(1-\lambda)\frac{J(B, C\overline{AB})}{J(B, C\overline{B})}, \\
&= \overline{B}(1-\lambda)\,{}_2\mathbb{F}_1\begin{bmatrix} \overline{A}C & B \\ & C \end{bmatrix}; \frac{\lambda}{\lambda-1}\end{bmatrix} + \delta(1-\lambda)\frac{J(B, C\overline{AB})}{J(B, C\overline{B})}, \\
&= \overline{AB}C(1-\lambda)\,{}_2\mathbb{F}_1\begin{bmatrix} \overline{A}C & \overline{B}C \\ & C \end{bmatrix}; \lambda\end{bmatrix} + \delta(1-\lambda)\frac{J(B, C\overline{AB})}{J(B, C\overline{B})}.
\end{aligned}$$

Using Proposition 6.8 and Lemma 2.8, we obtain the following proposition, which gives the evaluations in the imprimitive (see Definition 4.3) cases.

PROPOSITION 8.6. *Suppose* $\lambda \neq 0$, *and* A, B, C *are nontrivial. Then,*

$$\begin{aligned}
{}_2\mathbb{F}_1\begin{bmatrix} \varepsilon & B \\ & C \end{bmatrix}; \lambda\end{bmatrix} &= 1 - \frac{\overline{C}(\lambda)\overline{B}C(\lambda-1)}{J(B, \overline{B}C)}, \\
{}_2\mathbb{F}_1\begin{bmatrix} A & B \\ & B \end{bmatrix}; \lambda\end{bmatrix} &= \overline{A}(1-\lambda) - \overline{B}(\lambda)J(B, \overline{A}), \\
{}_2\mathbb{F}_1\begin{bmatrix} A & B \\ & A \end{bmatrix}; \lambda\end{bmatrix} &= \overline{B}(1-\lambda) - \frac{\overline{A}(\lambda)}{J(\overline{A}, B)}, \\
{}_2\mathbb{F}_1\begin{bmatrix} A & \varepsilon \\ & C \end{bmatrix}; \lambda\end{bmatrix} &= 1 - \overline{C}(-\lambda)\overline{A}C(1-\lambda)J(C, \overline{A}) - \delta(1-\lambda)\delta(\overline{A}C)(q-1).
\end{aligned}$$

Next, we consider the primitive (see Definition 4.3) cases.

PROPOSITION 8.7. *If* $A, B, C \in \widehat{\mathbb{F}_q^\times}$, $A, B \neq \varepsilon$ *and* $A, B \neq C$, *then we have the following.*

(1) *In general, we have*

$$J(A, \overline{A}C) \cdot {}_2\mathbb{P}_1\begin{bmatrix} A & B \\ & C \end{bmatrix}; \lambda\end{bmatrix} = J(B, \overline{B}C) \cdot {}_2\mathbb{P}_1\begin{bmatrix} B & A \\ & C \end{bmatrix}; \lambda\end{bmatrix},$$

$${}_2\mathbb{F}_1\begin{bmatrix} A & B \\ & C \end{bmatrix}; \lambda\end{bmatrix} = {}_2\mathbb{F}_1\begin{bmatrix} B & A \\ & C \end{bmatrix}; \lambda\end{bmatrix};$$

(2) *For* $\lambda \neq 0, 1$, *we have*

$${}_2\mathbb{P}_1\begin{bmatrix} A & B \\ & C \end{bmatrix}; \lambda\end{bmatrix} = \overline{C}(\lambda)C\overline{AB}(\lambda-1)\frac{J(B, C\overline{B})}{J(A, C\overline{A})}\,{}_2\mathbb{P}_1\begin{bmatrix} \overline{A} & B \\ & \overline{C} \end{bmatrix}; \lambda\end{bmatrix},$$

$${}_2\mathbb{F}_1\begin{bmatrix} A & B \\ & C \end{bmatrix}; \lambda\end{bmatrix} = \overline{C}(\lambda)C\overline{AB}(\lambda-1)\frac{J(\overline{B}, \overline{C}B)}{J(A, C\overline{A})}\,{}_2\mathbb{F}_1\begin{bmatrix} \overline{A} & B \\ & \overline{C} \end{bmatrix}; \lambda\end{bmatrix}.$$

PROOF. Since A, B are not equal to ε, C, part (1) follows from the definition of the $_2\mathbb{P}_1$ function (see (4.5)) and the relations $J(A, C\overline{A}) = A(-1)J(A, \overline{C})$, and $g(\overline{C}A)J(A, \overline{C}) = g(A)g(\overline{C})$.

To prove part (2), we first use the Kummer relations stated in Propositions 8.1 and 8.4. More precisely, for $\lambda \neq 0, 1$, we have

$$_2\mathbb{P}_1\begin{bmatrix} A & B \\ & C \end{bmatrix}; \lambda \overset{\text{Prop.8.1,part 3}}{=} B(-1)\,_2\mathbb{P}_1\begin{bmatrix} A & B \\ & AB\overline{C} \end{bmatrix}; 1-\lambda$$

$$\overset{\text{Prop.8.4,part 3}}{=} B(-1)\overline{C}(\lambda)\,_2\mathbb{P}_1\begin{bmatrix} B\overline{C} & A\overline{C} \\ & AB\overline{C} \end{bmatrix}; 1-\lambda$$

$$\overset{\text{Prop.8.1,part 3}}{=} \overline{C}(\lambda)ABC(-1)\,_2\mathbb{P}_1\begin{bmatrix} B\overline{C} & A\overline{C} \\ & \overline{C} \end{bmatrix}; \lambda$$

$$\overset{\text{Prop.8.4,part 3}}{=} \overline{C}(\lambda)\overline{AB}C(\lambda-1)\,_2\mathbb{P}_1\begin{bmatrix} \overline{B} & \overline{A} \\ & \overline{C} \end{bmatrix}; \lambda.$$

Next by part (1), one has

$$_2\mathbb{P}_1\begin{bmatrix} A & B \\ & C \end{bmatrix}; \lambda = \overline{C}(\lambda)\overline{AB}C(\lambda-1)\frac{J(\overline{A}, \overline{C}A)}{J(\overline{B}, \overline{C}B)}\,_2\mathbb{P}_1\begin{bmatrix} \overline{A} & \overline{B} \\ & \overline{C} \end{bmatrix}; \lambda.$$

$\square$

REMARK 8.8. Using the Galois perspective, one can interpret the above equalities as in Remark 8.3. Let $a, b, c \in \mathbb{Q}$, with $a, b, a-c, b-c \notin \mathbb{Z}$ and $\lambda \in \mathbb{Q}$. Let N be the least common denominator of a, b and c. Then for each prime ℓ, by part (2) of the above proposition, the 2-dimensional ℓ-adic Galois representation $\sigma_{\lambda,\ell}$ of $G_{\mathbb{Q}(\zeta_N)}$ associated with $_2P_1\begin{bmatrix} a & b \\ & c \end{bmatrix}; \lambda$ via Theorem 1.1 satisfies the following property: up to semisimplification,

$$\sigma_{\lambda,\ell} \cong \psi \otimes \overline{\sigma}_{\lambda,\ell},$$

where $\overline{\sigma}_{\lambda,\ell}$ is the complex conjugation of $\sigma_{\lambda,\ell}$ and ψ is the linear representation of G_K associated with the Grössencharacter $\chi_{-c,\lambda} \cdot \chi_{c-a-b,\lambda-1} \cdot \mathcal{J}_{(b,c-b)}/\mathcal{J}_{(a,c-a)}$ of K; see (5.5) and (5.8) for $\chi_{(\cdot),(\cdot)}$ and $\mathcal{J}_{(\cdot,\cdot)}$.

The next result corresponds to the classical equation (3.10) in §3.2.3.

COROLLARY 8.9. Suppose $A, B \neq \varepsilon$, and $A, B \neq C$. If $\lambda \neq 0$, we have

$$2 \cdot {}_2\mathbb{P}_1\begin{bmatrix} A & B \\ & C \end{bmatrix}; \lambda = ABC(-1)\overline{A}(\lambda)\,_2\mathbb{P}_1\begin{bmatrix} A & A\overline{C} \\ & A\overline{B} \end{bmatrix}; \frac{1}{\lambda}$$

$$+ ABC(-1)\overline{B}(\lambda)\frac{J(B, \overline{B}C)}{J(A, \overline{A}C)}\,_2\mathbb{P}_1\begin{bmatrix} B & B\overline{C} \\ & B\overline{A} \end{bmatrix}; \frac{1}{\lambda},$$

$$2 \cdot {}_2\mathbb{F}_1 \begin{bmatrix} A & B \\ & C \end{bmatrix}; \lambda \end{bmatrix} = ABC(-1)\overline{A}(\lambda)\frac{J(\overline{C}A, \overline{B}C)}{J(B, C\overline{B})} {}_2\mathbb{F}_1 \begin{bmatrix} A & A\overline{C} \\ & A\overline{B} \end{bmatrix}; \frac{1}{\lambda} \end{bmatrix}$$

$$+ ABC(-1)\overline{B}(\lambda)\frac{J(C\overline{A}, \overline{C}B)}{J(A, \overline{A}C)} {}_2\mathbb{F}_1 \begin{bmatrix} B & B\overline{C} \\ & B\overline{A} \end{bmatrix}; \frac{1}{\lambda} \end{bmatrix}.$$

Note that the appearance of a factor of 2 on the left hand sides corresponds to the fact that the right hand sides are the traces of 4-dimensional Galois representations at Frobenius elements.

PROOF. Using Proposition 8.7 part (1) and then Proposition 8.1 part (2) we obtain that if $\lambda \neq 0$,

$${}_2\mathbb{P}_1 \begin{bmatrix} A & B \\ & C \end{bmatrix}; \lambda \end{bmatrix} = \frac{J(B, \overline{B}C)}{J(A, C\overline{A})} {}_2\mathbb{P}_1 \begin{bmatrix} B & A \\ & C \end{bmatrix}; \lambda \end{bmatrix}$$

$$= ABC(-1)\overline{B}(\lambda)\frac{J(B, \overline{B}C)}{J(A, C\overline{A})} {}_2\mathbb{P}_1 \begin{bmatrix} B & \overline{C}B \\ & \overline{A}B \end{bmatrix}; \frac{1}{\lambda} \end{bmatrix}.$$

Combining this with the result obtained by using the second part of Proposition 8.1 directly, we get the desired relations. □

8.2. A Pfaff-Saalschütz evaluation formula

In this section we review an analogue of the Pfaff-Saalschütz formula obtained by Greene in [**41**], and provide geometric interpretations in terms of Galois representations.

To review how Greene obtained his analogue, we first relabel the Pfaff-Saalschütz formula (3.15) as

$$(8.2) \qquad {}_3F_2 \begin{bmatrix} a & b & -n \\ d & 1+a+b-n-d \end{bmatrix}; 1 \end{bmatrix} = \frac{(d-a)_n(d-b)_n}{(d)_n(d-a-b)_n}.$$

Next, take A, B and C to be $\overline{A}D, \overline{B}D$ and D in the third identity of Prop. 8.4 to obtain

$$(8.3) \quad AB\overline{D}(1-\lambda) {}_2\mathbb{P}_1 \begin{bmatrix} A & B \\ & D \end{bmatrix}; \lambda \end{bmatrix} = {}_2\mathbb{P}_1 \begin{bmatrix} \overline{A}D & \overline{B}D \\ & D \end{bmatrix}; \lambda \end{bmatrix} - \delta(1-\lambda)J(D\overline{B}, \overline{A})$$

where we use Lemma 2.8 to get the δ-term.

Now, expand the left side of (8.3) using the defining formula (4.4), invoking the rewritten third identity of Prop. 8.4 above, then expand the ${}_2\mathbb{P}_1$ on the right via its definition (4.5) to get a double sum. The δ-term becomes $BD(-1)J(D\overline{B}, \overline{A})$ and upon interchanging the order of summation one realizes the main term as $J(B, \overline{CB}D)J(C, A\overline{D})$. Finally, an application of Lemma 2.8 to $J(B, \overline{CB}D)$ establishes the following analogue to the Pfaff-Saalschütz formula (8.2),

$$(8.4) \quad {}_3\mathbb{P}_2 \begin{bmatrix} A & B & C \\ D & ABC\overline{D} \end{bmatrix}; 1 \end{bmatrix} = B(-1)J(C, A\overline{D})J(B, C\overline{D}) - BD(-1)J(D\overline{B}, \overline{A}).$$

Equivalently, for $a, b, c, d \in \mathbb{Q}$ with least common denominator N and any unramified prime ideal $\mathfrak{p}$ of $\mathcal{O}_{\mathbb{Q}(\zeta_N)}$, using $\mathcal{J}_{(a,b)}(\mathfrak{p}) = -\iota_\mathfrak{p}(a)\iota_\mathfrak{p}(b)(-1)J(\iota_\mathfrak{p}(a), \iota_\mathfrak{p}(b))$ as

(5.8) in §5.4 (See Definition 5.9 for $\iota_{\mathfrak{p}}(\cdot)$), one can rewrite the analogue (8.4) as

$$_3\mathbb{P}_2 \begin{bmatrix} \iota_{\mathfrak{p}}(a) & \iota_{\mathfrak{p}}(b) & \iota_{\mathfrak{p}}(c) \\ & \iota_{\mathfrak{p}}(d) & \iota_{\mathfrak{p}}(a+b+c-d) \end{bmatrix}; 1 \end{bmatrix}$$
$$= \chi_{a,-1}(\mathfrak{p}) \left[\mathcal{J}_{(c,a-d)}(\mathfrak{p}) \mathcal{J}_{(b,c-d)}(\mathfrak{p}) + \mathcal{J}_{(d-b,-a)}(\mathfrak{p}) \right].$$

In terms of Jacobi sums, (8.4) can be written as

$$(8.5) \quad \frac{1}{q-1} \sum_{\chi \in \widehat{\mathbb{F}_q^\times}} C\chi(-1) J(A\chi, \overline{\chi}) J(B\chi, \overline{D\chi}) J(C\chi, \overline{DABC\chi})$$
$$= J(C, A\overline{D}) J(B, C\overline{D}) - D(-1) J(D\overline{B}, \overline{A}).$$

REMARK 8.10. We note that from the Galois perspective, the $_3\mathbb{P}_2$ in (8.4) corresponds to a 2-dimensional Galois representation that can be described using two Grössencharacters. Furthermore, the additional term $-BD(-1)J(D\overline{B}, \overline{A})$ is due to the additional term involving $\delta(1-\lambda)$ appearing in the third identity in Proposition 8.4, together with formula (4.2).

For further examples of evaluation formulas over finite fields, see [**27, 41, 44, 70**].

8.3. A few analogues of algebraic hypergeometric formulas

In this section, we will give finite field analogues of (1.3) and the following identity [**88**, (1.5.19)]

$$(8.6) \quad _2F_1 \begin{bmatrix} a & a + \frac{1}{2} \\ & \frac{1}{2} \end{bmatrix}; z \end{bmatrix} = \frac{1}{2} \left((1+\sqrt{z})^{-2a} + (1-\sqrt{z})^{-2a} \right),$$

which generalize a few recent results of Tu and Yang in [**93**] that are proved using quotients of Fermat curves.

In the complex setting, to equate two formal power series, it suffices to compare the coefficients of each. For a proof of the identity (8.6) we note that both sides are functions of z. The coefficient of z^n on the right hand side is $\binom{-2a}{2n}$, while on the left it is

$$\frac{(a)_n (a+1/2)_n}{(1/2)_n (1)_n} \overset{\text{Thm. 2.5}}{=} \frac{(2a)_{2n}}{(1)_{2n}} = \binom{-2a}{2n}.$$

The evaluation identity (1.3) mentioned in the introduction can be proved similarly.

In the finite field setting, one can obtain identities in a similar way using (4.1) which we recall says that any function $f(x) : \mathbb{F}_q \to \mathbb{C}$ can be written uniquely in the form

$$f(x) = \delta(x) f(0) + \sum_{\chi \in \widehat{\mathbb{F}_q^\times}} f_\chi \cdot \chi(x).$$

We now state our finite field analogues of (8.6) and (1.3).

THEOREM 8.11. Let q be an odd prime power, $z \in \mathbb{F}_q^\times$, and $A \in \widehat{\mathbb{F}_q^\times}$ have order larger than 2. Then

$$_2\mathbb{F}_1 \begin{bmatrix} A & A\phi \\ & \phi \end{bmatrix}; z \end{bmatrix} = \left(\frac{1 + \phi(z)}{2} \right) \left(\overline{A}^2 (1+\sqrt{z}) + \overline{A}^2 (1-\sqrt{z}) \right),$$

$$_2\mathbb{F}_1 \begin{bmatrix} A & A\phi \\ & A^2 \end{bmatrix}; z \end{bmatrix} = \left(\frac{1 + \phi(1-z)}{2} \right) \left(\overline{A}^2 \left(\frac{1+\sqrt{1-z}}{2} \right) + \overline{A}^2 \left(\frac{1-\sqrt{1-z}}{2} \right) \right).$$

8.3. A FEW ANALOGUES OF ALGEBRAIC HYPERGEOMETRIC FORMULAS

REMARK 8.12. We first note that the above formulas are well-defined. When $z \neq 0$ does not have a square root in $\mathbb{F}_q$, then $\left(\frac{1+\phi(z)}{2}\right) = \frac{1-1}{2} = 0$. When $z \neq 0$ has a square root, we have $\left(\frac{1+\phi(z)}{2}\right) = 1$ and the right hand side is a sum of two characters.

REMARK 8.13. We now interpret the first identity in terms of a global Galois perspective. Assume $z \in \mathbb{Q}$ and $a = \frac{m}{N}$ for some integer m coprime to N. Let $\sigma_{z,\ell}$ be the 2-dimensional Galois representation associated with $_2F_1\begin{bmatrix} a & a+\frac{1}{2} \\ & \frac{1}{2} \end{bmatrix}; z$ as in Theorem 1.1. If $\sqrt{z} \notin \mathbb{Q}(\zeta_N)$, then for any unramified prime ideal $\mathfrak{p}$ of $\mathcal{O}_{\mathbb{Q}(\zeta_N)}$ that is inert in $\mathbb{Q}(\sqrt{z},\zeta_N)$, $\mathrm{Tr}\sigma_{z,\ell}(\mathrm{Frob}_\mathfrak{p}) = 0$. This means $\sigma_{z,\ell}$ is induced from a finite order character of $G_{\mathbb{Q}(\sqrt{z},\zeta_N)}$ which is an index-2 subgroup of $G_{\mathbb{Q}(\zeta_N)}$. There is a similar interpretation for the second result.

PROOF. By the first equality of Proposition 8.5, we have

$$_2F_1\begin{bmatrix} A & A\phi \\ & \phi \end{bmatrix}; z = \phi(z)\, _2F_1\begin{bmatrix} A & A\phi \\ & \phi \end{bmatrix}; z + \delta(z).$$

Hence, if z is not a square in $\mathbb{F}_q^\times$, the evaluation is equal to zero.

We now suppose $z = a^2$ for some $a \in \mathbb{F}_q^\times$. Then one has

$$_2F_1\begin{bmatrix} A & A\phi \\ & \phi \end{bmatrix}; z \stackrel{(4.7)}{=} \frac{A\phi(-1)}{(q-1)J(A\phi,\overline{A})} \sum_{\chi \in \widehat{\mathbb{F}_q^\times}} J(A\chi,\overline{\chi})J(A\phi\chi, \phi\overline{\chi})\chi(a^2)$$

$$\stackrel{(2.13)}{=} \frac{\phi A(-1)}{(q-1)J(A\phi,\overline{A})} \sum_\chi \frac{g(A\chi)g(\overline{\chi})}{g(A)} \frac{g(A\phi\chi)g(\phi\overline{\chi})}{g(A)} \chi(a^2)$$

$$\stackrel{(2.12)}{=} \frac{\phi(-1)}{q-1} \sum_\chi \frac{\overline{A\chi}(4)g(A^2\chi^2)g(\phi)^2 g(\overline{\chi}^2)\chi(4)}{q\overline{A}(4)g(A^2)} \chi(a^2)$$

$$\stackrel{(2.10)\text{ and }(2.13)}{=} \frac{1}{q-1} \sum_\chi \left(J(A^2\chi^2, \overline{\chi}^2) - (q-1)\delta(A^2) \right) \chi(a^2)$$

$$= \overline{A}^2(1+a) + \overline{A}^2(1-a).$$

We can use Proposition 8.5 and the duplication formula (2.12) to deduce that

$$_2F_1\begin{bmatrix} A & A\phi \\ & A^2 \end{bmatrix}; z = \frac{J(\phi, \phi A)}{J(\phi A, \phi A)}\, _2F_1\begin{bmatrix} A & A\phi \\ & \phi \end{bmatrix}; 1-z$$

$$= A(4)\, _2F_1\begin{bmatrix} A & A\phi \\ & \phi \end{bmatrix}; 1-z,$$

which leads to the desired second statement. □

Now we recast some of the previous discussion in terms of representations. Recall equation (1.3) mentioned in the introduction, which says

$$_2F_1\begin{bmatrix} a-\frac{1}{2} & a \\ & 2a \end{bmatrix}; z = \left(\frac{1+\sqrt{1-z}}{2}\right)^{1-2a}.$$

Writing the rational number $1 - 2a$ in reduced form, let N be its denominator. We first notice that when z is viewed as an indeterminate, the Galois group G

of the Galois closure of $\overline{\mathbb{Q}}\left(\left(\frac{1+\sqrt{1-z}}{2}\right)^{1-2a}\right)$ over $\overline{\mathbb{Q}}(z)$ is a Dihedral group. This is compatible with the corresponding monodromy group, which can be computed from the method explained in §3.2.4.

As a consequence of the second part of Theorem 8.11, we have the following corollary.

COROLLARY 8.14. *Let* $A, B \in \widehat{\mathbb{F}_q^\times}$ *such that* $A, B, AB, A\overline{B}$ *have orders larger than 2. Then,*

$$_2\mathbb{F}_1\begin{bmatrix} A & \phi A \\ & A^2 \end{bmatrix}; z\end{bmatrix}\,_2\mathbb{F}_1\begin{bmatrix} B & \phi B \\ & B^2 \end{bmatrix}; z\end{bmatrix}$$
$$= {}_2\mathbb{F}_1\begin{bmatrix} AB & \phi AB \\ & (AB)^2 \end{bmatrix}; z\end{bmatrix} + \overline{B}^2\left(\frac{z}{4}\right)\,_2\mathbb{F}_1\begin{bmatrix} A\overline{B} & \phi A\overline{B} \\ & (A\overline{B})^2 \end{bmatrix}; z\end{bmatrix} - \delta(1-z)AB(4),$$

$$_2\mathbb{F}_1\begin{bmatrix} A & \phi A \\ & \phi \end{bmatrix}; z\end{bmatrix}\,_2\mathbb{F}_1\begin{bmatrix} B & \phi B \\ & \phi \end{bmatrix}; z\end{bmatrix}$$
$$= {}_2\mathbb{F}_1\begin{bmatrix} AB & \phi AB \\ & \phi \end{bmatrix}; z\end{bmatrix} + \overline{B}^2(1-z)\,_2\mathbb{F}_1\begin{bmatrix} A\overline{B} & \phi A\overline{B} \\ & \phi \end{bmatrix}; z\end{bmatrix} - \delta(z).$$

Geometrically, this means the tensor product of two 2-dimensional representations of a dihedral group corresponding to the left side above is isomorphic to the direct sum of the two 2-dimensional representations corresponding to the terms on the right side of the above identities. In other words, it describes a fusion rule for 2-dimensional representations of Dihedral groups.

PROOF. Note that $\left(\frac{1+\sqrt{1-z}}{2}\right)\left(\frac{1-\sqrt{1-z}}{2}\right) = \frac{z}{4}$ and

$$\left(\frac{1+\phi(1-z)}{2}\right)^2 = \left(1 - \frac{\delta(1-z)}{2}\right)\left(\frac{1+\phi(1-z)}{2}\right).$$

For any element $z \neq 1 \in \mathbb{F}_q$, we have

$$_2\mathbb{F}_1\begin{bmatrix} A & \phi A \\ & A^2 \end{bmatrix}; z\end{bmatrix}\,_2\mathbb{F}_1\begin{bmatrix} B & \phi B \\ & B^2 \end{bmatrix}; z\end{bmatrix} =$$

$$\left(\frac{1+\phi(1-z)}{2}\right) \cdot \left(\overline{A}^2\left(\frac{1+\sqrt{1-z}}{2}\right) + \overline{A}^2\left(\frac{1-\sqrt{1-z}}{2}\right)\right)$$
$$\cdot \left(\overline{B}^2\left(\frac{1+\sqrt{1-z}}{2}\right) + \overline{B}^2\left(\frac{1-\sqrt{1-z}}{2}\right)\right)$$
$$= \left(\frac{1+\phi(1-z)}{2}\right)\left(\overline{A}^2\overline{B}^2\left(\frac{1+\sqrt{1-z}}{2}\right) + \overline{A}^2\overline{B}^2\left(\frac{1-\sqrt{1-z}}{2}\right)\right)$$
$$+ \overline{B}^2\left(\frac{z}{4}\right)\left(\frac{1+\phi(1-z)}{2}\right)\left(\overline{A}^2 B^2\left(\frac{1+\sqrt{1-z}}{2}\right) + \overline{A}^2 B^2\left(\frac{1-\sqrt{1-z}}{2}\right)\right)$$
$$= {}_2\mathbb{F}_1\begin{bmatrix} AB & \phi AB \\ & (AB)^2 \end{bmatrix}; z\end{bmatrix} + \overline{B}^2\left(\frac{z}{4}\right)\,_2\mathbb{F}_1\begin{bmatrix} A\overline{B} & \phi A\overline{B} \\ & (A\overline{B})^2 \end{bmatrix}; z\end{bmatrix}.$$

8.3. A FEW ANALOGUES OF ALGEBRAIC HYPERGEOMETRIC FORMULAS

When $z = 1$,
$$_2\mathbb{F}_1\begin{bmatrix} \chi & \phi\chi \\ & \chi^2 \end{bmatrix}; 1\end{bmatrix} = \frac{1}{2} \cdot 2\overline{\chi}^2\left(\frac{1}{2}\right) = \chi(4),$$
for any character χ. Therefore,
$$_2\mathbb{F}_1\begin{bmatrix} A & \phi A \\ & A^2 \end{bmatrix}; 1\end{bmatrix} {}_2\mathbb{F}_1\begin{bmatrix} B & \phi B \\ & B^2 \end{bmatrix}; 1\end{bmatrix} = A(4)B(4)$$
$$= \frac{1}{2}\left({}_2\mathbb{F}_1\begin{bmatrix} AB & \phi AB \\ & (AB)^2 \end{bmatrix}; 1\end{bmatrix} + \overline{B}^2\left(\frac{1}{4}\right) {}_2\mathbb{F}_1\begin{bmatrix} A\overline{B} & \phi A\overline{B} \\ & (A\overline{B})^2 \end{bmatrix}; 1\end{bmatrix}\right).$$

The first statement follows.

The second statement follows in a way similar to the first, using the first part of Theorem 8.11 or the transformation
$$_2\mathbb{F}_1\begin{bmatrix} A & \phi A \\ & A^2 \end{bmatrix}; z\end{bmatrix} = A(4) \,{}_2\mathbb{F}_1\begin{bmatrix} A & \phi AB \\ & \phi \end{bmatrix}; 1-z\end{bmatrix}$$
used in the proof of Theorem 8.11. $\square$

Additionally, consider Slater's equation (1.5.21) in [**88**],
$$_2F_1\begin{bmatrix} 2a & a+1 \\ & a \end{bmatrix}; z\end{bmatrix} = \frac{1+z}{(1-z)^{2a+1}}.$$
We have the following finite field analogue: for $A \neq \varepsilon$,
$$_2\mathbb{F}_1\begin{bmatrix} A^2 & A \\ & A \end{bmatrix}; z\end{bmatrix} = \overline{A}^2(1-z) - \overline{A}(z)J(A, \overline{A}^2).$$
This follows from the second part of Proposition 8.6.

REMARK 8.15. The two formulas stated in Theorem 8.11 are direct consequences of the duplication formula in Theorem 2.5. By Theorem 2.6, we can generalize (8.6) using the same argument to get the following formula representing the mth multiplication formula

$$(8.7) \quad {}_m F_{m-1}\begin{bmatrix} a & a+\frac{1}{m} & \cdots & a+\frac{m-1}{m} \\ & \frac{1}{m} & \cdots & \frac{m-1}{m} \end{bmatrix}; z\end{bmatrix} = \frac{1}{m}\left(\sum_{i=1}^{m}(1 - \zeta_m^i \sqrt[m]{z})^{-ma}\right).$$

Geometrically, the monodromy group for the degree-m differential equation satisfied by the function on the left is a finite group (see the criterion in [**17**]), which is isomorphic to the Galois group of $\overline{\mathbb{Q}}((1 + \sqrt[m]{z})^{-ma})$ over the function field $\overline{\mathbb{Q}}(z)$. When $m = 2$ it is a Dihedral group. In general, it is a cyclic group extended by $\mathbb{Z}/m\mathbb{Z}$. Its finite field analogue is of the following form: let $q \equiv 1 \pmod{m}$ and η_m be a primitive order m character, then for $A \in \widehat{\mathbb{F}_q^\times}$ such that $A^m \neq \varepsilon$,

$$(8.8) \quad {}_m\mathbb{F}_{m-1}\begin{bmatrix} A & A\eta_m & \cdots & A\eta_m^{m-1} \\ & \eta_m & \cdots & \eta_m^{m-1} \end{bmatrix}; z\end{bmatrix}$$
$$= \frac{1}{m}\left(1 + \sum_{i=1}^{m-1}\eta_m(z)\right)\sum_{i=1}^{m}\overline{A}^m\left(1 - \zeta_m^i \sqrt[m]{z}\right).$$

Equation ((8.8) follows from the definition of $_m\mathbb{F}_{m-1}$ and the multiplication formula for Jacobi sums (2.20).

We conclude this section by continuing a discussion about $g(t)^2 = {}_2F_1\left[\begin{smallmatrix}\frac{1}{4} & \frac{3}{4} \\ & \frac{2}{3}\end{smallmatrix}; t\right]^2$ in Example 3.7.

EXAMPLE 8.16. By the Clausen formula (7.1),

$$g(t)^2 \stackrel{(3.8)}{=} (1-t)^{-\frac{1}{2}} {}_2F_1\left[\begin{matrix}\frac{1}{4} & -\frac{1}{12} \\ & \frac{2}{3}\end{matrix}; \frac{t}{t-1}\right]^2$$

$$\stackrel{(7.1)}{=} (1-t)^{-\frac{1}{2}} {}_3F_2\left[\begin{matrix}-\frac{1}{6} & \frac{1}{6} & \frac{1}{2} \\ & \frac{2}{3} & \frac{1}{3}\end{matrix}; \frac{t}{t-1}\right]$$

$$\stackrel{(8.7)}{=} \frac{(1-t)^{-\frac{1}{2}}}{3}\left(\sum_{i=1}^{3}\sqrt{1 - \zeta_3^i \sqrt[3]{\frac{t}{t-1}}}\right).$$

For the relation between $f(t)^2 = {}_2F_1\left[\begin{smallmatrix}\frac{1}{4} & \frac{3}{4} \\ & \frac{4}{3}\end{smallmatrix}; t\right]^2$ and $g(t)^2$, if we let $x := x(t) = \sqrt{3 + 6g(t)^2}$, then

$$f(t)^2 = -8 - \frac{4x(x+1)(x-3)}{3}\cdot\frac{t-1}{t}$$
$$= 8g(t)^2(2 - \sqrt{3 + 6g(t)^2})\frac{t-1}{t} - \frac{8}{t}.$$

Now we consider the $\mathbb{F}_q$ analogue with argument $t = \frac{\lambda}{\lambda-1}$ so that the results are easier to state. Let $q \equiv 1 \pmod{12}$ be a prime power and η be a primitive order 12 character on $\mathbb{F}_q^\times$. Then the finite field versions of $f\left(\frac{\lambda}{\lambda-1}\right)^2$ and $g\left(\frac{\lambda}{\lambda-1}\right)^2$ can be stated as follows: for $\lambda \neq 0, 1$,

(8.9) $\eta^8(\lambda)\, {}_2\mathbb{F}_1\left[\begin{matrix}\eta^3 & \overline{\eta}^3 \\ & \eta^4\end{matrix}; \frac{\lambda}{\lambda-1}\right]^2 = {}_2\mathbb{F}_1\left[\begin{matrix}\eta^3 & \overline{\eta}^3 \\ & \overline{\eta}^4\end{matrix}; \frac{\lambda}{\lambda-1}\right]^2$

$$= \begin{cases} 1 + \sum_{0 \le i < j \le 2} \eta^6\left((1-\zeta_3^i a)(1-\zeta_3^j a)\right), & \text{if } \lambda = a^3 \\ \eta^4(\lambda), & \text{otherwise.} \end{cases}$$

This is obtained by using the following finite field Pfaff transformation given in item 4 of Proposition 8.5,

$${}_2\mathbb{F}_1\left[\begin{matrix}\eta^3 & \overline{\eta}^3 \\ & \eta^4\end{matrix}; \frac{\lambda}{\lambda-1}\right]^2 = \eta^6(1-\lambda)\, {}_2\mathbb{F}_1\left[\begin{matrix}\eta^3 & \overline{\eta} \\ & \eta^4\end{matrix}; \lambda\right]^2,$$

the finite field Clausen formula (Theorem 7.1 with $S = \eta^3, C = \overline{\eta}^4$)

$${}_2\mathbb{F}_1\left[\begin{matrix}\eta^3 & \overline{\eta} \\ & \eta^4\end{matrix}; \lambda\right]^2 = {}_3\mathbb{F}_2\left[\begin{matrix}\overline{\eta}^2 & \eta^2 & \eta^6 \\ & \eta^4 & \overline{\eta}^4\end{matrix}; \lambda\right] + \eta^6(1-\lambda)\eta^4(\lambda), \quad \lambda \neq 1,$$

equation (8.8) which gives that

$${}_3\mathbb{F}_2\left[\begin{matrix}\overline{\eta}^2 & \eta^2 & \eta^6 \\ & \eta^4 & \overline{\eta}^4\end{matrix}; \lambda\right] = \frac{1}{3}(1 + \eta^4(\lambda) + \overline{\eta}^4(\lambda))\left(\sum_{i=0}^{2}\eta^6\left(1 - \zeta_3^i \sqrt[3]{\lambda}\right)\right),$$

the last identity of Proposition 8.7,
$$_2\mathbb{F}_1\left[\begin{matrix} \eta^3 & \overline{\eta}^3 \\ & \overline{\eta}^4 \end{matrix}; \frac{\lambda}{\lambda-1}\right] = \eta^4(\lambda)\frac{J(\eta^5,\eta^3)}{J(\eta,\eta^3)} {}_2\mathbb{F}_1\left[\begin{matrix} \eta^3 & \overline{\eta}^3 \\ & \eta^4 \end{matrix}; \frac{\lambda}{\lambda-1}\right],$$
and the fact $J(\eta^5,\eta^3) = J(\eta,\eta^3)$ when η has order 12.

CHAPTER 9

Quadratic or Higher Transformation Formulas

In this chapter we first consider finite field analogues of some higher degree hypergeometric transformation formulas which are related to elliptic curves. These highlight the role geometric correspondences like isogeny and isomorphism play in some transformation formulas.

Next, we obtain several finite field analogues of classical formulas satisfying the ($*$) condition given on page 1 by using our main technique. Our first example is a quadratic formula, which demonstrates that our technique has the capacity to produce analogues that are satisfied by all values in $\mathbb{F}_q$. We prove analogues of the Bailey cubic $_3F_2$ formulas and an analogue of a formula by Andrews and Stanton.

We then use a different approach to obtain a finite field analogue of a cubic formula of Gessel and Stanton. As a corollary of this cubic formula, Gessel and Stanton obtained an evaluation formula, with a proof that cannot be translated directly. However, using the Galois perspective we can predict a finite field analogue which we then prove using a different approach. This application illustrates that the Galois perspective is helpful in a greater context.

9.1. Some results related to elliptic curves

In §8.1, we discussed the finite field analogues of Kummer's 24 relations, which are between $_2\mathbb{P}_1$ (resp. $_2\mathbb{F}_1$) functions linked via linear fractional transformations. These are the cases that can be deduced from Greene's finite field version of the Lagrange theorem stated in Theorem 2.10. However, Greene's theorem does not apply to higher degree transformation formulas.

Borwein and Borwein ([**18**]) proved that for real $z \in (0,1)$,

$$_2F_1\left[\begin{array}{cc} \frac{1}{3} & \frac{2}{3} \\ & 1 \end{array}; 1-z^3\right] = \frac{3}{1+2z} \,_2F_1\left[\begin{array}{cc} \frac{1}{3} & \frac{2}{3} \\ & 1 \end{array}; \left(\frac{1-z}{1+2z}\right)^3\right].$$

Geometrically, this corresponds to the fact that the two elliptic curves

$$y^2 + xy + \frac{1-z^3}{27}y = x^3, \qquad y^2 + xy + \frac{1}{27}\frac{(1-z)^3}{(1+2z)^3}y = x^3$$

are 3-isogenous over $\mathbb{Q}(z, \zeta_3)$, which can be verified using the degree-3 modular equation satisfied by their corresponding j-invariants [**25**]

$$X^4 + Y^4 + 36864000(X^3 + Y^3) + 452984832000000(X^2 + Y^2)$$
$$+ 1855425871872000000000(X + Y) - X^3Y^3 + 2587918086X^2Y^2$$
$$- 770845966336000000 XY + 2332(X^3Y^2 + X^2Y^3)$$
$$- 1069956(X^3Y + XY^3) + 8900222976000(X^2Y + XY^2).$$

Hence, when $q \equiv 1 \pmod{3}$ and z can be embedded in $\mathbb{F}_q$, the local zeta functions of these curves over $\mathbb{F}_q$ are the same, i.e. if η_3 is a primitive cubic character in $\widehat{\mathbb{F}_q^\times}$, then

$$\text{(9.1)} \qquad {}_2\mathbb{P}_1\left[\begin{matrix} \eta_3 & \eta_3^2 \\ & \varepsilon \end{matrix}; 1-z^3\right] = {}_2\mathbb{P}_1\left[\begin{matrix} \eta_3 & \eta_3^2 \\ & \varepsilon \end{matrix}; \left(\frac{1-z}{1+2z}\right)^3\right].$$

When one replaces ${}_2\mathbb{P}_1$ by ${}_2\mathbb{F}_1$, the above formula still holds by our definition of ${}_2\mathbb{F}_1$ (4.8) as the normalizing Jacobi sums are the same.

Similarly, we consider the relation between the universal elliptic curves

$$E_t : y^2 = 4x^3 - \frac{27x}{1-t} - \frac{27}{1-t},$$

and the Legendre curves

$$L_\lambda : y^2 = x(1-x)(1-\lambda x).$$

The j-invariant of E_t is $\frac{1728}{t}$ and the j-invariant of L_λ is $\frac{256(\lambda^2-\lambda+1)^3}{\lambda^2(\lambda-1)^2}$. Stienstra and Beukers ([**89**, Theorem 1.5]) proved

$$\text{(9.2)} \qquad {}_2F_1\left[\begin{matrix} \frac{1}{12} & \frac{5}{12} \\ & 1 \end{matrix}; \frac{27\lambda^2(\lambda-1)^2}{4(\lambda^2-\lambda+1)^3}\right] = (1-\lambda+\lambda^2)^{1/4} \, {}_2F_1\left[\begin{matrix} \frac{1}{2} & \frac{1}{2} \\ & 1 \end{matrix}; \lambda\right].$$

Geometrically, L_λ is the pullback of the universal elliptic curve along the natural map of modular curves $Y(2) \to Y(1)$.

To see the finite field analogue of (9.2), we have the following relation by the work of the first author [**37**, Theorem 1.2].

THEOREM 9.1. *For $p \equiv 1 \pmod{12}$, $\lambda \in \mathbb{F}_p$ with $\lambda \neq -1, 2, 1/2$ and $\lambda^2 - \lambda + 1 \neq 0$, we have*

$${}_2\mathbb{P}_1\left[\begin{matrix} \eta_{12} & \eta_{12}^5 \\ & \varepsilon \end{matrix}; \frac{27\lambda^2(\lambda-1)^2}{4(\lambda^2-\lambda+1)^3}\right] = \eta_{12}(-1)\eta_{12}^3(1-\lambda+\lambda^2) \, {}_2\mathbb{P}_1\left[\begin{matrix} \phi & \phi \\ & \varepsilon \end{matrix}; \lambda\right],$$

where η_{12} is a primitive character of order 12 in $\widehat{\mathbb{F}_p^\times}$. To write it in terms of ${}_2F_1$,

$$\text{(9.3)} \qquad {}_2F_1\left[\begin{matrix} \eta_{12} & \eta_{12}^5 \\ & \varepsilon \end{matrix}; \frac{27\lambda^2(\lambda^2-1)^2}{4(\lambda^2-\lambda+1)^3}\right] = \eta_{12}^3(1-\lambda+\lambda^2) \, {}_2F_1\left[\begin{matrix} \phi & \phi \\ & \varepsilon \end{matrix}; \lambda\right].$$

PROOF. For any fixed $\lambda \in \mathbb{Q} \setminus \{0,1\}$, the elliptic curve L_λ is isomorphic to E_t with $t = \frac{27\lambda^2(\lambda-1)^2}{4(\lambda^2-\lambda+1)^3}$ over $\mathbb{Q}$. To be more explicit, when $t = \frac{27\lambda^2(\lambda-1)^2}{4(\lambda^2-\lambda+1)^3}$, the elliptic curve E_t can be written as

$$\frac{(\lambda-2)^2(2\lambda-1)^2(\lambda+1)^2}{4}y^2 =$$
$$\left((2\lambda^2+\lambda-1)x + 3(\lambda^2-\lambda+1)\right)\left((\lambda^2-\lambda-2)x - 3(\lambda^2-\lambda+1)\right)$$
$$\cdot \left((2\lambda^2-5\lambda+2)x + 3(\lambda^2-\lambda+1)\right).$$

It is $\mathbb{Q}$-isomorphic to

$$Y^2 = -\frac{\lambda^2-\lambda+1}{(\lambda-2)(\lambda+1)(2\lambda-1)}X(X-1)(X-(1-\lambda))$$

by the change of variables

$$(X,Y) = \left(-\frac{3(\lambda^2-\lambda+1)}{2\lambda^2+\lambda-1}\left(\frac{3x}{\lambda-2}+1\right), \frac{(\lambda-2)(2\lambda-1)(\lambda+1)}{2^4(\lambda^2-\lambda+1)}y\right).$$

For any prime $p \equiv 1 \pmod{12}$, let $a_p(t)$ denote the trace of the Frobenius endomorphism on E_t. According to the work of the first author [37], when $t \in \mathbb{F}_p^\times$ and $t \neq 1$, i.e. $\lambda \in \mathbb{F}_p^\times$ with $\lambda \neq \pm 1, 2, 1/2$, we have

$$_2\mathbb{P}_1\begin{bmatrix} \eta_{12} & \eta_{12}^5 \\ & \varepsilon \end{bmatrix}; t \end{bmatrix} = -\phi(2)\eta_{12}(-1)\overline{\eta}_{12}^3(1-t)a_p(t).$$

By the above discussion

$$a_p(t) = -\phi((1-\lambda+\lambda^2)(\lambda-2)(2\lambda-1))\,_2\mathbb{P}_1\begin{bmatrix} \phi & \phi \\ & \varepsilon \end{bmatrix}; \lambda \end{bmatrix}.$$

It follows that when $\lambda \neq 0, \pm 1, 2, 1/2$ in $\mathbb{F}_p$,

$$_2\mathbb{P}_1\begin{bmatrix} \eta_{12} & \eta_{12}^5 \\ & \varepsilon \end{bmatrix}; \frac{27\lambda^2(\lambda-1)^2}{4(\lambda^2-\lambda+1)^3} \end{bmatrix} = \eta_{12}(-1)\eta_{12}^3(1-\lambda+\lambda^2)\,_2\mathbb{P}_1\begin{bmatrix} \phi & \phi \\ & \varepsilon \end{bmatrix}; \lambda \end{bmatrix}.$$

One can check this identity also holds for $\lambda = 0, 1$. We deduce (9.3) by noting that

$$J(\eta_{12}^5, \eta_{12}^7) = \sum_{x \in \mathbb{F}_q} \eta_{12}^5(x)\eta_{12}^7(1-x) = \sum_{x \in \mathbb{F}_q \setminus \{1\}} \eta_{12}^5\left(\frac{x}{1-x}\right) = -\eta_{12}^5(-1)$$

as stated in (2.14). Similarly, $J(\phi, \phi) = -\phi(-1)$. □

We wish to keep in mind that geometrically, transformation formulas are sometimes related to 'correspondences' like isogenies or isomorphisms, however in general the underlying geometric objects are complicated. More abstractly, these transformation formulas describe 'correspondences' between two hypergeometric motives. From this perspective, finite field analogues of these transformation formulas describe relations between the Galois representations associated with the motives. The two examples above demonstrate that when the involved monodromy groups are arithmetic and hence have moduli interpretations, one may obtain $\mathbb{F}_q$ transformation formulas geometrically. The techniques used later in this section are mainly based on character sums and they can handle general cases regardless of whether there exist moduli interpretations.

9.2. A Kummer quadratic transformation formula

In this section, we illustrate our dictionary method with a finite field analogue of Kummer's quadratic formula. Recall from §3.2.7 the Kummer quadratic transformation formula

$$(9.4) \quad (1-x)^{-c}\,_2F_1\begin{bmatrix} \frac{1+c}{2}-b & \frac{c}{2} \\ & c-b+1 \end{bmatrix}; \frac{-4x}{(1-x)^2} \end{bmatrix} = \,_2F_1\begin{bmatrix} b & c \\ & c-b+1 \end{bmatrix}; x \end{bmatrix}.$$

We now outline a proof using the multiplication and reflection formulas, along with the Pfaff-Saalschütz formula. Observe that our proof is different from the one in [6, pp. 125-126]. We first recall an inversion formula [88, (2.2.3.1)].

$$(9.5) \qquad (a)_{n-r} = (-1)^r \frac{(a)_n}{(1-a-n)_r}.$$

9. QUADRATIC OR HIGHER TRANSFORMATION FORMULAS

PROOF. To begin, note that

$$(9.6) \quad \binom{-c-2k}{n-k} = (-1)^{n-k}\frac{(c+2k)_{n-k}}{(1)_{n-k}} \stackrel{(9.5)}{=} (-1)^{n-k}\frac{(c+2k)_n(-n)_k}{(1-c-2k-n)_k(1)_n}$$

$$= (-1)^{n-k}\frac{\Gamma(c+2k+n)\Gamma(1-c-2k-n)}{\Gamma(c+2k)\Gamma(1-c-k-n)}\frac{(-n)_k}{n!}$$

$$\stackrel{\text{reflection}}{=} (-1)^n\frac{\Gamma(c+2k+n)\Gamma(c+k+n)}{\Gamma(c+2k)\Gamma(c+2k+n)}\frac{(-n)_k}{n!} = (-1)^n\frac{(c)_{n+k}(-n)_k}{(c)_{2k}n!}$$

$$= (-1)^n\frac{(c)_n(c+n)_k(-n)_k}{(c)_{2k}n!} \stackrel{(2.8)}{=} (-1)^n\frac{(c)_n(c+n)_k(-n)_k}{4^k(\frac{c}{2})_k(\frac{c+1}{2})_k n!}.$$

The left hand side of (9.4) can be expanded as

$$\sum_{k\geq 0}\frac{(\frac{1+c}{2}-b)_k(\frac{c}{2})_k}{k!(c-b+1)_k}(-4x)^k(1-x)^{-c-2k}$$

$$= \sum_{k\geq 0}\frac{(\frac{1+c}{2}-b)_k(\frac{c}{2})_k}{k!(c-b+1)_k}(-4x)^k\sum_{i\geq 0}\binom{-c-2k}{i}(-x)^i$$

$$\stackrel{n=k+i}{=} \sum_{k,n\geq 0}\frac{(\frac{1+c}{2}-b)_k(\frac{c}{2})_k}{k!(c-b+1)_k}4^k\binom{-c-2k}{n-k}(-x)^n$$

$$\stackrel{(9.6)}{=} \sum_{k,n\geq 0}\frac{(\frac{1+c}{2}-b)_k(\frac{c}{2})_k}{k!(c-b+1)_k}4^k\frac{(c)_n(c+n)_k(-n)_k}{4^k(\frac{c}{2})_k(\frac{c+1}{2})_k n!}x^n$$

$$= \sum_{n\geq 0}\frac{(c)_n}{n!}\,{}_3F_2\!\left[\begin{matrix}\frac{1+c}{2}-b & c+n & -n \\ & \frac{c+1}{2} & c-b+1\end{matrix}; 1\right]x^n.$$

By the Pfaff-Saalschütz formula (3.15), the above equals

$$\sum_{n\geq 0}\frac{(c)_n(b)_n(\frac{1-c}{2}-n)_n}{n!(\frac{1+c}{2})_n(b-c-n)_n}x^n \stackrel{\text{reflection}}{=} \sum_{n\geq 0}\frac{(b)_n(c)_n}{n!(c-b+1)_n}x^n.$$

$\square$

Now we obtain the corresponding finite field analogue of (9.4), starting with an analogue of (9.6). We recall that ϕ denotes the quadratic character, and note that under the assumptions below, $\frac{c}{2}$ in the classical setting corresponds to a character D below.

LEMMA 9.2. Let $D, K, \chi \in \widehat{\mathbb{F}_q^\times}$ and set $C = D^2$. Then,

$$(9.7) \quad J(\overline{CK^2}, \overline{\chi}K) = \chi D\phi(-1)\frac{g(\overline{\chi}K)g(CK\chi)g(D)g(D\phi)}{K(4)g(DK)g(C)}\frac{g(\overline{DK\phi})}{q}$$

$$+ \frac{(q-1)^2}{q}\delta(D\chi\phi)\delta(DK\phi) + (q-1)\delta(DK).$$

9.2. A KUMMER QUADRATIC TRANSFORMATION FORMULA

PROOF.

$$J(\overline{CK^2}, \overline{\chi}K) \stackrel{\text{Lem.2.8}}{=} \chi K(-1) J(\overline{\chi}K, C\chi K)$$

$$= \chi K(-1) \frac{g(\overline{\chi}K) g(C\chi K)}{g(CK^2)} \cdot \frac{g(D^2)}{g(C)} + (q-1)\delta(CK^2)$$

$$\stackrel{\text{duplication}}{=} \chi K(-1) \frac{g(\overline{\chi}K) g(C\chi K) g(D) g(D\phi)}{K(4) g(DK) g(DK\phi) g(C)} + (q-1)\delta(CK^2)$$

$$\stackrel{(2.11)}{=} \chi K(-1) \frac{g(\overline{\chi}K) g(C\chi K) g(D) g(D\phi)}{K(4) g(DK) g(C)}$$

$$\cdot \left(\frac{DK\phi(-1)}{q} g(\overline{DK\phi}) - \frac{q-1}{q} \delta(DK\phi) \right) + (q-1)\delta(CK^2).$$

Breaking this into three terms, the middle term is

$$- \chi K(-1) \frac{g(\overline{\chi}K) g(C\chi K) g(D) g(D\phi)}{K(4) g(DK) g(C)} \frac{(q-1)}{q} \delta(DK\phi)$$

$$\stackrel{\text{duplication}}{=} -\chi K(-1) \frac{(q-1)}{q} g(\overline{\chi D}\phi) g(\chi D\phi) \delta(D\phi K)$$

$$\stackrel{(2.10)}{=} -(q-1)\delta(DK\phi) + \frac{(q-1)^2}{q} \delta(D\phi K) \delta(D\phi\chi).$$

Recombining the three terms, using the easily established identity $\delta(R^2) - \delta(R\phi) = \delta(R)$, gives the Lemma. $\square$

REMARK 9.3. Among the three terms on the right hand side of (9.7), the first is the major term predicted by the classical case. The other two correspond to degenerate cases.

THEOREM 9.4. *Let* $B, D \in \widehat{\mathbb{F}_q^\times}$, *and set* $C = D^2$. *When* $D \neq \phi$ *and* $B \neq D$, *we have, for all* $x \in \mathbb{F}_q$

$$\overline{C}(1-x) \,_2F_1 \begin{bmatrix} D\phi\overline{B} & D \\ & C\overline{B} \end{bmatrix} ; \frac{-4x}{(1-x)^2} \end{bmatrix}$$

$$= \,_2F_1 \begin{bmatrix} B & C \\ & C\overline{B} \end{bmatrix} ; x \end{bmatrix} - \delta(1-x) \frac{J(C, \overline{B}^2)}{J(C, \overline{B})} - \delta(1+x) \frac{J(\overline{B}, D\phi)}{J(C, \overline{B})}.$$

REMARK 9.5. In the above statement, we see the character $\overline{C}(1-x)$ multiplied by a hypergometric function being evaluated at a rational expression which is undefined when $x = 1$. In this instance (and others similar to it), we take the convention that the value of the product is 0 when $x = 1$.

We will use the following identities, which can be checked directly, in the proof of Theorem 9.4. When $B \neq C$, and $\phi D \overline{B} \neq \varepsilon$,

$$(9.8) \qquad \overline{D}(4) \frac{J(\phi\overline{B}, D)}{J(D, D\overline{B})} = \frac{J(C, \overline{B}^2)}{J(C, \overline{B})}; \quad \frac{J(\overline{B}, D\phi)}{J(C, \overline{B})} = \frac{J(C\overline{B}, D\phi)}{J(C, D\overline{B}\phi)}.$$

PROOF. When $x = 0$, both sides are equal to 1 by definition. When $x = 1$, both sides take value 0 as well. Thus, we assume $x \neq 0, 1$.

We first consider the case when $C = D^2 \neq \varepsilon$, and $B^2 \neq C$. We deal with this case by considering separately when $B = C$ and when $B \neq C$. Suppose $B \neq C$. Taking $A_1 = D\phi\overline{B}$, $A_2 = D$, $B_1 = C\overline{B}$ and $K = \chi$ in (4.7),

$$\overline{C}(1-x) \, _2F_1 \begin{bmatrix} D\phi\overline{B} & D \\ & C\overline{B} \end{bmatrix} ; \frac{-4x}{(1-x)^2} \end{bmatrix}$$

$$= \frac{1}{(q-1)J(D, D\overline{B})} \sum_{K \in \widehat{\mathbb{F}_q^\times}} D(-1) J(D\phi\overline{B}K, \overline{K}) J(DK, B\overline{C}\overline{K}) K(-4x) \overline{CK^2}(1-x)$$

$$= \frac{1}{(q-1)^2 J(D, D\overline{B})} \sum_{K \in \widehat{\mathbb{F}_q^\times}} D(-1) J(D\phi\overline{B}K, \overline{K}) J(DK, B\overline{C}\overline{K})$$

$$\cdot K(-4x) \sum_{\varphi \in \widehat{\mathbb{F}_q^\times}} J(\overline{CK^2}, \overline{\varphi}) \varphi(x).$$

Letting $\chi := K\varphi$, the above equals

$$(9.9) \quad \frac{1}{(q-1)^2 J(D, D\overline{B})}$$
$$\cdot \sum_{K, \chi \in \widehat{\mathbb{F}_q^\times}} D(-1) J(D\phi\overline{B}K, \overline{K}) J(DK, B\overline{C}\overline{K}) K(-4) J(\overline{CK^2}, \overline{\chi}K) \chi(x).$$

By Lemma 9.2 and (9.8) we can rewrite this as

$$(9.10) \quad S_1 + S_2 - \overline{D}(4) \frac{J(\phi\overline{B}, D)}{J(D, D\overline{B})} \delta(1-x) = S_1 + S_2 - \delta(1-x) \frac{J(C, \overline{B}^2)}{J(C, \overline{B})},$$

where the S_i are obtained by applying (9.7). Explicitly,

$$S_1 = \frac{1}{(q-1)^2 J(D, D\overline{B})} \sum_{K, \chi \in \widehat{\mathbb{F}_q^\times}} D(-1) J(D\phi\overline{B}K, \overline{K}) J(DK, B\overline{C}\overline{K}) K(-4)$$

$$\cdot \chi D\phi(-1) \frac{g(\overline{\chi}K) g(CK\chi) g(D) g(D\phi)}{K(4) g(DK) g(C)} \frac{g(\overline{DK\phi})}{q} \chi(x);$$

and

$$S_2 = \frac{\overline{D}(4) J(\phi, \phi B\overline{D})}{q J(D, D\overline{B})} \phi(-1) J(\overline{B}, \phi D) \phi \overline{D}(x).$$

9.2. A KUMMER QUADRATIC TRANSFORMATION FORMULA

We first analyze S_1.

$$S_1 = \frac{g(C\overline{B})}{(q-1)^2 g(D)g(D\overline{B})} \sum_{K,\chi \in \widehat{\mathbb{F}_q^\times}} \chi\phi(-1) J(D\phi\overline{B}K, \overline{K}) \frac{g(DK)g(B\overline{CK})}{g(B\overline{D})}$$

$$\cdot K(-4) \frac{g(\overline{\chi}K)g(CK\chi)g(D)g(D\phi)}{K(4)g(DK)g(C)} \frac{g(\overline{DK}\phi)}{q} \chi(x)$$

$$\stackrel{\text{combine Gauss sums}}{=} \frac{BD(-1)g(C\overline{B})g(D\phi)}{(q-1)^2 q^2 g(C)} \sum_\chi g(B\overline{C\chi})g(D\phi\chi)$$

$$\cdot \left(\sum_K K\chi\phi(-1) J(D\phi\overline{B}K, \overline{K}) \frac{g(B\overline{CK})g(\overline{\chi}K)}{g(B\overline{C\chi})} \frac{g(CK\chi)g(\overline{DK}\phi)}{g(D\phi\chi)} \right) \chi(x).$$

Letting $\Omega := \frac{BD(-1)g(C\overline{B})g(D\phi)}{q^2 g(C)}$, we have

$$S_1 = \frac{\Omega}{(q-1)^2} \sum_{\chi \in \widehat{\mathbb{F}_q^\times}} g(B\overline{C\chi})g(D\phi\chi)$$

$$\cdot \left(\sum_{K \in \widehat{\mathbb{F}_q^\times}} K\chi\phi(-1) J(D\phi\overline{B}K, \overline{K}) \frac{g(B\overline{CK})g(\overline{\chi}K)}{g(B\overline{C\chi})} \frac{g(CK\chi)g(\overline{DK}\phi)}{g(D\phi\chi)} \right) \chi(x)$$

$$\stackrel{\text{Jacobi sums}}{=} \frac{\Omega}{(q-1)^2} \sum_\chi \phi(-1) g(B\overline{C\chi})g(D\phi\chi)\chi(x) \sum_K K\chi(-1) J(D\phi\overline{B}K, \overline{K})$$

$$\cdot \left[J(B\overline{CK}, \overline{\chi}K) - (q-1) BCK(-1) \delta(B\overline{C\chi}) \right]$$

$$\cdot \left[J(CK\chi, \overline{DK}\phi) - (q-1) DK\phi(-1) \delta(D\phi\chi) \right].$$

To continue we separate the delta terms,

$$S_1 = \frac{\Omega}{(q-1)} \sum_\chi \phi(-1) g(B\overline{C\chi})g(D\phi\chi)\chi(x) \cdot$$

$$\sum_K \frac{K\chi(-1)}{q-1} J(D\phi\overline{B}K, \overline{K}) J(B\overline{CK}, \overline{\chi}K) J(CK\chi, \overline{DK}\phi) + E_1 + E_2$$

$$\stackrel{(8.5)}{=} \frac{\Omega}{(q-1)} \sum_\chi \phi(-1) g(B\overline{C\chi})g(D\phi\chi)\chi(x)$$

$$\cdot \left(J(C\chi, \overline{D}\phi) J(\overline{\chi}, B\chi) - B(-1) J(C\overline{B}\chi, \overline{D}\phi B) \right) + E_1 + E_2,$$

where

$$E_1 := -\sum_{\chi \in \widehat{\mathbb{F}_q^\times}} \frac{\Omega\phi(-1)}{(q-1)} g(\varepsilon) g(B\overline{D}\phi) \sum_{K \in \widehat{\mathbb{F}_q^\times}} J(D\phi\overline{B}K, \overline{K}) J(BK, \overline{DK}\phi) \cdot \delta(B\overline{C\chi}) \chi(x)$$

$$\stackrel{\text{Gauss eval.}}{=} \sum_\chi \Omega\phi(-1) g(B\overline{D}\phi) J(B, \overline{B}) \delta(B\overline{C\chi}) \chi(x)$$

$$= -\phi B(-1) \Omega g(B\overline{D}\phi) B\overline{C}(x),$$

$$E_2 := -\sum_{\chi \in \widehat{\mathbb{F}_q^\times}} \frac{\Omega \phi(-1)}{(q-1)} g(\varepsilon) g(B\overline{D}\phi) \sum_{K \in \widehat{\mathbb{F}_q^\times}} J(D\phi \overline{B}K, \overline{K}) J(B\overline{CK}, D\phi K) \delta(D\phi\chi) \chi(x)$$

$$\stackrel{\text{Gauss eval.}}{=} \sum_{\chi} \frac{g(C\overline{B}) g(D\phi)}{q^2 g(C)} \phi(-1) g(B\overline{D}\phi) J(D\phi, \overline{D}\phi) \delta(D\phi\chi) \chi(x)$$

$$= -D(-1) \Omega g(B\overline{D}\phi) \overline{D}\phi(x).$$

Now we continue our analysis of S_1. We separate $S_1 - (E_1 + E_2)$ into two terms. The first term is

$$\frac{\Omega}{(q-1)} \sum_{\chi \in \widehat{\mathbb{F}_q^\times}} \phi(-1) g(B\overline{C}\overline{\chi}) g(D\phi\chi) J(C\chi, \overline{D}\phi) J(\overline{\chi}, B\chi) \chi(x)$$

$$= \frac{\Omega}{(q-1)} \sum_{\chi} \phi(-1) g(B\overline{C}\overline{\chi}) g(D\phi\chi)$$

$$\cdot \left(\frac{g(C\chi) g(\overline{D}\phi)}{g(D\phi\chi)} + (q-1) D\phi(-1) \delta(D\phi\chi) \right) J(\overline{\chi}, B\chi) \chi(x).$$

By definition, this equals

$$_2\mathbb{F}_1 \begin{bmatrix} B & C \\ & C\overline{B} \end{bmatrix}; x \Bigg] - \sum_{\chi \in \widehat{\mathbb{F}_q^\times}} \Omega g(B\overline{D}\phi) D(-1) J(D\phi, B\overline{D}\phi) \delta(D\phi\chi) \chi(x)$$

$$= {_2\mathbb{F}_1} \begin{bmatrix} B & C \\ & C\overline{B} \end{bmatrix}; x \Bigg] - \Omega g(B\overline{D}\phi) \phi(-1) J(D\phi, \overline{B}) \overline{D}\phi(x)$$

$$= {_2\mathbb{F}_1} \begin{bmatrix} B & C \\ & C\overline{B} \end{bmatrix}; x \Bigg] - S_2.$$

The second term of $S_1 - (E_1 + E_2)$ is

$$-\frac{\Omega}{(q-1)} \sum_{\chi \in \widehat{\mathbb{F}_q^\times}} \phi(-1) g(B\overline{C}\overline{\chi}) g(D\phi\chi) B(-1) J(C\overline{B}\chi, \overline{D}\phi B) \chi(x)$$

$$= -\frac{\Omega}{(q-1)} \sum_{\chi} \phi(-1) g(B\overline{C}\overline{\chi}) g(D\phi\chi) B(-1) CB\chi(-1) J(C\overline{B}\chi, \overline{D}\overline{\chi}\phi) \chi(x)$$

$$= -\frac{\Omega}{(q-1)} \sum_{\chi} \phi\chi(-1) g(B\overline{C}\overline{\chi}) g(D\phi\chi) \frac{g(C\overline{B}\chi) g(\overline{D}\overline{\chi}\phi)}{g(D\overline{B}\phi)} \chi(x)$$

$$= -\frac{\Omega}{(q-1)} \sum_{\chi} \frac{\phi\chi(-1)}{g(D\overline{B}\phi)} g(B\overline{C}\overline{\chi}) g(C\overline{B}\chi) g(D\phi\chi) g(\overline{D}\overline{\chi}\phi) \chi(x).$$

9.2. A KUMMER QUADRATIC TRANSFORMATION FORMULA

Applying the finite field reflection formula (2.10), the above becomes

$$-\frac{\Omega}{(q-1)}\sum_{\chi\in\widehat{\mathbb{F}_q^\times}}\frac{\phi\chi(-1)}{g(D\overline{B}\phi)}\left(BC\chi(-1)q-(q-1)\delta(B\overline{C}\chi)\right)$$

$$\cdot (D\phi\chi(-1)q-(q-1)\delta(D\phi\chi))\chi(x)$$

$$=-\frac{q^2\Omega}{(q-1)g(D\overline{B}\phi)}DB(-1)\sum_\chi \chi(-1)\chi(x) - E_1 - E_2$$

$$\stackrel{(4.2)}{=} -\frac{q^2\Omega}{g(D\overline{B}\phi)}DB(-1)\delta(1+x) - E_1 - E_2$$

$$= -\delta(1+x)\frac{J(\overline{B},D\phi)}{J(C,\overline{B})} - E_1 - E_2.$$

Thus we have that

$$S_1 = {}_2F_1\begin{bmatrix} B & C \\ & C\overline{B} \end{bmatrix}; x\bigg] - \delta(1+x)\frac{J(\overline{B},D\phi)}{J(C,\overline{B})} - S_2.$$

Thus by (9.10), putting these all together gives that when $B \neq C$,

$$\overline{C}(1-x)\,{}_2F_1\begin{bmatrix} D\phi\overline{B} & D \\ & C\overline{B} \end{bmatrix}; \frac{-4x}{(1-x)^2}\bigg] =$$

$$\quad {}_2F_1\begin{bmatrix} B & C \\ & C\overline{B} \end{bmatrix}; x\bigg] - \delta(1-x)\frac{J(C,\overline{B}^2)}{J(C,\overline{B})} - \delta(1+x)\frac{J(\overline{B},D\phi)}{J(C,\overline{B})},$$

as desired.

When $B = C$, we can obtain by similar means that

$$\overline{C}(1-x)\,{}_2F_1\begin{bmatrix} \phi\overline{D} & D \\ & \varepsilon \end{bmatrix}; \frac{-4x}{(1-x)^2}\bigg]$$

$$= {}_2F_1\begin{bmatrix} C & C \\ & \varepsilon \end{bmatrix}; x\bigg] - \delta(1-x)\frac{J(C,\overline{C}^2)}{J(C,\overline{C})} - \delta(1+x)\frac{J(\overline{C},D\phi)}{J(C,\overline{C})}.$$

We next consider the case when $D^2 = \varepsilon$. Since we know $D \neq \phi$ from our hypothesis, we have $D = \varepsilon$. Note that under the assumption of $B \neq D$, we also have $B \neq \varepsilon$. In this case, by Proposition 8.6, the left hand side of the equation in

Theorem 9.4 is

$$\varepsilon(1-x) \, {}_2\mathbb{F}_1 \left[\begin{matrix} \phi\overline{B} & \varepsilon \\ & \overline{B} \end{matrix} ; \frac{-4x}{(1-x)^2} \right]$$

$$= - \left(B\left(\frac{4x}{(1-x)^2}\right) \phi\left(\frac{(1+x)^2}{(1-x)^2}\right) J(\overline{B}, B\phi) - \varepsilon(1-x) \right)$$

$$= -B\left(\frac{4x}{(1-x)^2}\right) J(\overline{B}, B\phi)(1 - \delta(1+x)) + 1$$

$$= -B\left(\frac{4x}{(1-x)^2}\right) J(\overline{B}, B\phi) + 1 + B(-1) J(\overline{B}, B\phi)\delta(1+x)$$

$$= -B\left(\frac{4x}{(1-x)^2}\right) J(\overline{B}, B\phi) + 1 + J(\overline{B}, \phi)\delta(1+x)$$

$$= -B\left(\frac{4x}{(1-x)^2}\right) J(\overline{B}, B\phi) + 1 - \frac{J(\overline{B}, \phi)}{J(\varepsilon, B)} \delta(1+x).$$

Moreover, by Proposition 8.6, and Lemma 2.8 we have that

$$ {}_2\mathbb{F}_1 \left[\begin{matrix} B & \varepsilon \\ & \overline{B} \end{matrix} ; x \right] = - \left(B(-x) \overline{B}^2 (1-x) J(\overline{B}, \overline{B}) - 1 \right)$$

$$\stackrel{(2.16)}{=} -B\left(\frac{-4x}{(1-x)^2}\right) J(\overline{B}, \phi) + 1 = -B\left(\frac{4x}{(1-x)^2}\right) J(\overline{B}, B\phi) + 1.$$

Recalling that $\delta(1-x) = 0$ since $x \neq 1$, this gives the desired result that

$$\varepsilon(1-x) \, {}_2\mathbb{F}_1 \left[\begin{matrix} \phi\overline{B} & \varepsilon \\ & \overline{B} \end{matrix} ; \frac{-4x}{(1-x)^2} \right]$$

$$= {}_2\mathbb{F}_1 \left[\begin{matrix} B & \varepsilon \\ & \overline{B} \end{matrix} ; x \right] - \delta(1-x) \frac{J(\varepsilon, \overline{B}^2)}{J(\varepsilon, \overline{B})} - \delta(1+x) \frac{J(\overline{B}, \phi)}{J(\varepsilon, \overline{B})}.$$

We now consider the case when $B^2 = C$, and thus $B = D\phi$. The left hand side becomes

$$\overline{C}(1-x) \, {}_2\mathbb{F}_1 \left[\begin{matrix} \varepsilon & D \\ & \overline{D\phi} \end{matrix} ; \frac{-4x}{(1-x)^2} \right] = \overline{C}(1-x) - \frac{\overline{D\phi}(-4x)}{J(D, \phi)}$$

$$= \overline{C}(x-1) - \frac{\overline{D\phi}(-4x)}{J(D, \phi)} (1 - \delta(1+x)),$$

while the right hand side is

$$\overline{C}(x-1) - \frac{\overline{D\phi}(x)}{J(C, \overline{D\phi})} = \overline{C}(x-1) - \frac{\overline{D\phi}(x)}{J(\overline{D\phi}, \overline{D\phi})}.$$

Using the duplication formula, we obtain the result. □

9.2.1. A summary of the the proof of Theorem 9.4. In the previous proof of the quadratic formula, the delta terms produce two kinds of byproducts: the extra terms like $\delta(1 \pm x)$, which are due to the degeneracy of the corresponding Galois representations at the corresponding x values, and the extra characters like the E_i terms described above. While the first kind of extra terms are unavoidable, but can be hidden by placing extra assumptions on the values of x, the second kind of extra terms might be eliminated. In particular, some of them are 'path dependent' in the following sense. In the primitive case by Definition (4.8) and Proposition 8.7, the $_2F_1$ function can be computed from two $_2\mathbb{P}_1$ functions whose upper parameters are related by a permutation. For instance, in this proof, if we swap the upper characters of the left hand side, then the delta terms do not involve $B\overline{C}(x)$.

The above approach can be adopted to give finite field analogues of many classical quadratic or higher degree formulas relating two primitive $_{n+1}F_n$ functions with rational parameters and arguments of the form $c_1 x$ and $c_2 x^{n_1}(1-c_3 x)^{n_2}$, respectively, that satisfies the $(*)$ condition, where $n_1, n_2 \in \mathbb{Z}$, $n_1 > 0$ and $c_1, c_2, c_3 \in \mathbb{Q}^\times$. See [10] for a collection of such formulas. As we assume the classical formula satisfies the $(*)$ condition, we will similarly mimic the proof of the classical formula to get the finite field analogues. We expand one side (typically the more complicated side) of the formula as a double summation involving two characters K and χ, as in (9.9). Then we reduce the double sum to obtain the other side of the formula. The dictionary described in §2.4 predicts that the major terms for the finite field analogues, i.e. the $_{n+1}\mathbb{F}_n$ terms predicted on both sides, agree. Though extra delta terms are technical to compute in general, they correspond to the discrepancy between two finite dimensional Galois representations. There is one subtlety we should be aware of in the finite field translation: we might need to extend the current finite field $\mathbb{F}_q$ in order to use an evaluation formula to bridge both sides.

9.3. The quadratic formula in connection with the Kummer relations

In this section, we give another finite field version of Kummer's quadratic formula in which the arguments on both sides of the transformation are rational functions of z of degree two.

Together with the Kummer relations discussed in §8.1, one can also obtain many equivalent versions of quadratic formulas over finite fields which are useful in many ways. A different approach to eliminate the extra characters of the second kind as described in §9.2.1 is as follows. Recall D is chosen so that $D^2 = C$. Applying a Pfaff transformation (the fifth identity of Proposition 8.5) to the right hand side of the quadratic formula in Theorem 9.4, we obtain

$$(9.11) \quad \overline{C}(1-x) \,_2\mathbb{F}_1 \begin{bmatrix} D\phi\overline{B} & D \\ & C\overline{B} \end{bmatrix};\, \frac{-4x}{(1-x)^2} \end{bmatrix}$$

$$= \overline{C}(1-x) \,_2\mathbb{F}_1 \begin{bmatrix} C\overline{B}^2 & C \\ & C\overline{B} \end{bmatrix};\, \frac{x}{x-1} \end{bmatrix} - \delta(1+x)\frac{J(\overline{B},D\phi)}{J(C,\overline{B})}.$$

By letting $z = \frac{x}{x-1}$ and hence $1-z = \frac{1}{1-x}$, when $z \neq \frac{1}{2}, 1$, we get

$$(9.12) \quad _2\mathbb{F}_1 \begin{bmatrix} D\phi\overline{B} & D \\ & C\overline{B} \end{bmatrix};\, 4z(1-z) \end{bmatrix} = \,_2\mathbb{F}_1 \begin{bmatrix} C\overline{B}^2 & C \\ & C\overline{B} \end{bmatrix};\, z \end{bmatrix},$$

or equivalently when also A^2, $B^2 \neq \varepsilon$, then

$$(9.13) \quad _2\mathbb{F}_1 \begin{bmatrix} A & B \\ & AB\phi \end{bmatrix};\, 4z(1-z) \end{bmatrix} = \,_2\mathbb{F}_1 \begin{bmatrix} A^2 & B^2 \\ & AB\phi \end{bmatrix};\, z \end{bmatrix}.$$

9. QUADRATIC OR HIGHER TRANSFORMATION FORMULAS

For example, it is easy to see that the terms do not differ by extra characters as both sides are invariant under the involution $z \mapsto 1 - z$.

Below is the $\mathbb{F}_q$ version of another Kummer quadratic transformation formula.

THEOREM 9.6. *Let $A, B \in \widehat{\mathbb{F}_q^\times}$ such that $B, B^2\overline{A}, A\overline{B}\phi \neq \varepsilon$. Then we have*

$$_2F_1 \begin{bmatrix} A & B \\ & B^2 \end{bmatrix}; \frac{4z}{(1+z)^2} \end{bmatrix} = A^2(1+z)\, _2F_1 \begin{bmatrix} A & A\phi\overline{B} \\ & B\phi \end{bmatrix}; z^2 \end{bmatrix},$$

when $z \neq \pm 1$.

This is the finite field analogue of the following quadratic transformation of Kummer (see [**6**, (3.1.11)])

$$_2F_1 \begin{bmatrix} a & b \\ & 2b \end{bmatrix}; \frac{4z}{(1+z)^2} \end{bmatrix} = (1+z)^{2a}\, _2F_1 \begin{bmatrix} a & a+1/2-b \\ & b+1/2 \end{bmatrix}; z^2 \end{bmatrix}.$$

PROOF OF THEOREM 9.6. Following arguments similar to the proof of the classical case, one can derive Kummer's quadratic transformation from the quadratic formulas

$$(9.14) \quad _2F_1 \begin{bmatrix} B & D^2 \\ & D^2\overline{B} \end{bmatrix}; z \end{bmatrix} = \overline{D}^2(1+z)\, _2F_1 \begin{bmatrix} \phi D & D \\ & D^2\overline{B} \end{bmatrix}; \frac{4z}{(z+1)^2} \end{bmatrix}, \quad z \neq \pm 1,$$

provided $B \neq \varepsilon$ and $D^2 \neq B^2$, and
(9.15)
$$_2F_1 \begin{bmatrix} K & E^2 \\ & K^2 \end{bmatrix}; 1-z \end{bmatrix} = E^2(2)\overline{E}^2(1+z)\, _2F_1 \begin{bmatrix} \phi E & E \\ & \phi K \end{bmatrix}; \frac{(z-1)^2}{(z+1)^2} \end{bmatrix}, \quad z \neq \pm 1,$$

provided $K \neq \varepsilon$ and $E^2 \neq K^2$.

The first equation is obtained by applying Pfaff's transformation (Proposition 8.5 - item 4) to the quadratic formula in Theorem 9.4. Using Proposition 8.5 - item 3 and the duplication formula on Gauss sums, one can observe the second identity from (9.14).

Replace $1 - z$ with $4z/(1+z)^2$ in (9.15) to derive the formula

$$_2F_1 \begin{bmatrix} K & E^2 \\ & K^2 \end{bmatrix}; \frac{4z}{(1+z)^2} \end{bmatrix} = \overline{E}^2(1+z^2)E^4(1+z)\, _2F_1 \begin{bmatrix} \phi E & E \\ & \phi K \end{bmatrix}; \left(\frac{2z}{1+z^2}\right)^2 \end{bmatrix}.$$

Replace the arguments $\frac{4z}{(1+z)^2}$ by $\left(\frac{2z}{1+z^2}\right)^2$ and set the characters $D = E$, $B = \phi E^2 \overline{K}$ in (9.14) to obtain the formula

$$_2F_1 \begin{bmatrix} \phi E^2 \overline{K} & E^2 \\ & \phi K \end{bmatrix}; z^2 \end{bmatrix} = \overline{E}^2(1+z^2)\, _2F_1 \begin{bmatrix} \phi E & E \\ & \phi K \end{bmatrix}; \left(\frac{2z}{1+z^2}\right)^2 \end{bmatrix}.$$

Therefore, one has

$$_2F_1 \begin{bmatrix} E^2 & K \\ & K^2 \end{bmatrix}; \frac{4z}{(1+z)^2} \end{bmatrix} = E^4(1+z)\, _2F_1 \begin{bmatrix} E^2 & \phi E^2 \overline{K} \\ & \phi K \end{bmatrix}; z^2 \end{bmatrix},$$

since the $_2F_1$-functions are primitive under our assumptions. One can remove the condition on the finite field $\mathbb{F}_q$ and replace the character E^2 by any character A. □

As an immediate corollary of Theorem 9.6, we obtain the finite field version of Theorem 3 in [**94**], which says for real numbers a and b such that neither $b + 3/4$ nor $2b + 1/2$ is a non-positive integer, then in a neighborhood of $z = 0$ one has

$$_2F_1\begin{bmatrix} a+b & b+\frac{1}{4} \\ & 2b+\frac{1}{2} \end{bmatrix}; \frac{4z}{(1+z)^2} \end{bmatrix} = (1+z)^{2a+2b}\, _2F_1\begin{bmatrix} a+b & a+\frac{1}{4} \\ & b+\frac{3}{4} \end{bmatrix}; z^2 \end{bmatrix}.$$

COROLLARY 9.7. *Let η be a primitive character of order 4 so $\phi = \eta^2$. Let A, B characters so that $A\eta$, $B\eta$, $A\overline{B}\phi \neq \varepsilon$. Then we have*

$$_2F_1\begin{bmatrix} AB & B\eta \\ & \phi B^2 \end{bmatrix}; \frac{4z}{(1+z)^2} \end{bmatrix} = A^2B^2(1+z)\, _2F_1\begin{bmatrix} AB & A\eta \\ & B\overline{\eta} \end{bmatrix}; z^2 \end{bmatrix},$$

when $z \neq 0, \pm 1$.

See Appendix §11.1 for some numeric observations related to other higher transformation formulas in [**94**].

9.3.1. Two other immediate corollaries of Theorem 9.4.
From a transformation formula, one can obtain evaluation formulas either by specifying values or by comparing coefficients on both sides. Here we will show how to obtain the finite field analogue of the Kummer evaluation formula (3.14), which has been obtained by Greene in [**41**] using a different approach.

COROLLARY 9.8. *Assume $B, C \in \widehat{\mathbb{F}_q^\times}$ and set $C = D^2$, then*

$$_2F_1\begin{bmatrix} B & C \\ & C\overline{B} \end{bmatrix}; -1 \end{bmatrix} = \frac{J(D, \overline{B})}{J(C, \overline{B})} + \frac{J(\overline{B}, D\phi)}{J(C, \overline{B})}.$$

PROOF. By Theorem 9.4, if $C = D^2 \neq \varepsilon$, $B^2 \neq C$, $x = -1$, one has

$$\overline{C}(2)\, _2F_1\begin{bmatrix} D\phi\overline{B} & D \\ & C\overline{B} \end{bmatrix}; 1 \end{bmatrix} = \, _2F_1\begin{bmatrix} B & C \\ & C\overline{B} \end{bmatrix}; -1 \end{bmatrix} - \frac{J(\overline{B}, D\phi)}{J(C, \overline{B})}.$$

Hence,

$$_2F_1\begin{bmatrix} B & C \\ & C\overline{B} \end{bmatrix}; -1 \end{bmatrix} = \overline{C}(2)\, _2F_1\begin{bmatrix} D\phi\overline{B} & D \\ & C\overline{B} \end{bmatrix}; 1 \end{bmatrix} + \frac{J(\overline{B}, D\phi)}{J(C, \overline{B})}$$

$$= \overline{D}(4)\frac{J(\phi, D)}{J(D, D\overline{B})} + \frac{J(\overline{B}, D\phi)}{J(C, \overline{B})},$$

where if $C \neq B$,

$$\overline{D}(4)\frac{J(\phi, D)}{J(D, D\overline{B})} = \frac{\overline{D}(4)g(\phi)g(D)g(C\overline{B})}{g(\phi D)g(D)g(D\overline{B})} = \frac{J(\overline{B}, D)}{J(C, \overline{B})},$$

and if $C = B$,

$$\overline{D}(4)\frac{J(\phi, D)}{J(D, D\overline{B})} = \frac{\overline{D}(4)J(\phi, D)}{J(D, \overline{D})} = \frac{J(D, D)}{J(D, \overline{D})} = \frac{J(\overline{C}, D)}{J(C, \overline{C})},$$

since $J(\phi, \chi) = \chi(4)J(\chi, \chi)$ if $\chi \neq \varepsilon$.

When $C = \varepsilon$, the claim follows Proposition 6.8. □

Next we will get a 'quadratic' version of the finite field Pfaff-Saalschütz formula (8.5) to be used later. This version includes cases to which (8.5) is not applicable. Switching x and $1-x$, we first rewrite the formula in Theorem 9.4 as

$$(9.16) \quad \overline{C}(x) \, _2F_1\left[\begin{matrix} D\phi\overline{B} & D \\ & C\overline{B} \end{matrix}; \frac{-4(1-x)}{x^2}\right] = {}_2F_1\left[\begin{matrix} B & C \\ & C\overline{B} \end{matrix}; 1-x\right] \\ - \delta(x)\frac{J(C,\overline{B}^2)}{J(C,\overline{B})} - \delta(2-x)\frac{J(\overline{B},D\phi)}{J(C,\overline{B})}.$$

We first use one of the Kummer relations to write the first term on the right in terms of x and assume $x \neq 0$. Now we look at the left hand side which is

$$\overline{C}(x) \, _2F_1\left[\begin{matrix} D\phi\overline{B} & D \\ & C\overline{B} \end{matrix}; \frac{-4(1-x)}{x^2}\right]$$

$$\stackrel{(4.6)}{=} \frac{D(-1)}{(q-1)J(D,D\overline{B})} \sum_{K \in \widehat{\mathbb{F}_q^\times}} J(D\phi\overline{B}K, \overline{K})J(DK, B\overline{CK})K(-4)\overline{CK^2}(x)K(1-x)$$

$$+ \overline{C}(x)\delta\left(\frac{-4(1-x)}{x^2}\right)$$

$$= \frac{D(-1)}{(q-1)^2 J(D,D\overline{B})} \sum_K J(D\phi\overline{B}K, \overline{K})J(DK, B\overline{CK})K(-4)\overline{CK^2}(x)$$

$$\cdot \sum_{\varphi \in \widehat{\mathbb{F}_q^\times}} J(K,\overline{\varphi})\varphi(x) + \overline{C}(x)\delta(1-x)$$

$$\stackrel{\chi=\overline{CK^2}\varphi}{=} \frac{D(-1)}{(q-1)^2 J(D,D\overline{B})} \sum_{\chi \in \widehat{\mathbb{F}_q^\times}} \sum_{K \in \widehat{\mathbb{F}_q^\times}} J(D\phi\overline{B}K, \overline{K})J(DK, B\overline{CK})K(-4)J(K,\overline{C\chi K^2})\chi(x)$$

$$+ \sum_{\chi \in \widehat{\mathbb{F}_q^\times}} \frac{1}{q-1}\chi(x),$$

where in the last step we use (4.2) to expand $\delta(1-x)$.

Since C is a square, $C(-1) = 1$. Under the assumption that $x \neq 0$, the right hand side of (9.16) is

$$_2F_1\left[\begin{matrix} B & C \\ & C\overline{B} \end{matrix}; 1-x\right] - \delta(2-x)\frac{J(\overline{B},D\phi)}{J(C,\overline{B})}$$

$$= \frac{J(C,\overline{B}^2)}{J(C,\overline{B})} \, _2F_1\left[\begin{matrix} B & C \\ & B^2 \end{matrix}; x\right] - \delta(x-2)\frac{J(\overline{B},D\phi)}{J(C,\overline{B})}$$

$$\stackrel{(4.2)}{=} \frac{1}{(q-1)J(C,\overline{B})}\left(\sum_{\chi \in \widehat{\mathbb{F}_q^\times}} J(B\chi,\overline{\chi})J(C\chi,\overline{B^2\chi})\chi(x) - J(\overline{B},D\phi)\sum_{\chi \in \widehat{\mathbb{F}_q^\times}}\overline{\chi}(2)\chi(x)\right).$$

By equating the coefficient of χ on both sides and assuming $B^2 \neq C$ we get

9.3. THE QUADRATIC FORMULA AND THE KUMMER RELATIONS

$$\left[\left(J(B\chi,\overline{\chi})J(C\chi,\overline{B^2\chi}) - J(\overline{B},D\phi)\overline{\chi}(2)\right)/J(C,\overline{B})\right] - 1$$

$$=\frac{D(-1)}{(q-1)J(D,D\overline{B})}\sum_{K\in\widehat{\mathbb{F}_q^\times}} J(D\phi\overline{B}K,\overline{K})J(DK,B\overline{CK})J(K,\overline{C\chi K^2})K(-4)$$

$$=\frac{D\chi(-1)}{(q-1)J(D,D\overline{B})}\sum_{K} J(D\phi\overline{B}K,\overline{K})J(DK,B\overline{CK})J(\overline{C\chi K^2},CK\chi)K(-4)$$

$$=\frac{D\chi(-1)}{(q-1)J(D,D\overline{B})}\sum_{K} \frac{g(D\phi\overline{B}K)g(\overline{K})g(DK)g(B\overline{CK})}{g(D\phi\overline{B})g(B\overline{D})}$$

$$\cdot \left(\frac{g(\overline{C\chi K^2})g(CK\chi)}{g(\overline{K})} + (q-1)\chi(-1)\delta(K)\right)K(-4)$$

$$=\left[\frac{D\chi(-1)}{(q-1)J(D,D\overline{B})}\sum_{K} \frac{g(D\phi\overline{B}K)g(DK)g(B\overline{CK})g(\overline{C\chi K^2})g(CK\chi)}{g(D\phi\overline{B})g(B\overline{D})}K(-4)\right] - 1.$$

Assume $C = D^2$, $D \neq \phi$ and $B^2 \neq C$. Then

(9.17) $\sum_{K\in\widehat{\mathbb{F}_q^\times}} g(D\phi\overline{B}K)g(DK)g(B\overline{CK})g(\overline{C\chi K^2})g(CK\chi)K(-4) =$

$(q-1)\dfrac{D\chi(-1)g(D\phi\overline{B})g(B\overline{D})J(D,D\overline{B})}{J(C,\overline{B})}\left(J(B\chi,\overline{\chi})J(C\chi,\overline{B^2\chi}) - J(\overline{B},D\phi)\overline{\chi}(2)\right).$

In conclusion, we obtained the following quadratic version of Pfaff-Saalschütz formula over $\mathbb{F}_q$.

COROLLARY 9.9. *Let A, B, $C \in \widehat{\mathbb{F}_q^\times}$ not simultaneously trivial, such that $CA, CB \neq \phi$. Then*

$$\frac{AB(-1)C(4)J(\phi AC,\overline{ABC^2})}{q(q-1)g(\phi)}\sum_{\chi\in\widehat{\mathbb{F}_q^\times}} g(A\chi^2)g(B\chi)g(C\overline{\chi})g(\phi\overline{ABC\chi})g(\overline{\chi})\overline{\chi}(-4)$$
$$= J(\overline{ABC^2},AC^2)J(A,B^2C^2) - J(\overline{ABC^2},\phi AC)AC^2(2).$$

If $CB = \phi$ (or $CA = \varepsilon$), then

$$\frac{1}{(q-1)g(\phi)}\sum_{\chi\in\widehat{\mathbb{F}_q^\times}} g(A\chi^2)g(B\chi)g(C\overline{\chi})g(\phi\overline{ABC\chi})g(\overline{\chi})\overline{\chi}(-4)$$
$$= qB(4)AB(-1)J(B,B\overline{A}) - qAB(-1)A(2) - \delta(A)(q-1)B(-1)J(B,\phi);$$

If $CA = \phi$ (or $CB = \varepsilon$), then

$$\frac{1}{(q-1)g(\phi)}\sum_{\chi\in\widehat{\mathbb{F}_q^\times}} g(A\chi^2)g(B\chi)g(\phi\overline{A\chi})g(\overline{B\chi})g(\overline{\chi})\overline{\chi}(-4)$$
$$= -B(4)J(\phi,B)J(A\overline{B}^2,\phi B\overline{A}) - qAB(-1)A(2)$$
$$+ \delta(A)q(q-1)B(-1) - \delta(\phi B)(q-1)AB(-1).$$

PROOF. The first case can be obtained from replacing the character K by $KC\overline{\chi}$ in equation (9.17) and relabeling the characters. The other cases are straightforward verifications from the relation between Gauss sums and Jacobi sums. □

REMARK 9.10. When A is a square of another character, the left hand side of the formulas in the above Corollary can be evaluated using the finite field Pfaff-SaalSchütz formula (8.5). However, one does not need this assumption in the above Corollary.

9.4. A finite field analogue of a theorem of Andrews and Stanton

In this section, we will prove a finite field analogue of the following transformation formula given by Andrews and Stanton [**7**]:

THEOREM 9.11 (Theorem 5 of [**7**]). *For $a, b, c, x \in \mathbb{C}$ such that both sides converge,*

$$(1-x)^{a+1} {}_3F_2 \left[\begin{matrix} a+1 & b+1 & a-b+\tfrac{1}{2} \\ & 2b+2 & 2a-2b+1 \end{matrix} ; 4x(1-x) \right]$$
$$= (1-x)^{-a-1} {}_3F_2 \left[\begin{matrix} a+1 & b+1 & a-b+\tfrac{1}{2} \\ & 2b+2 & 2a-2b+1 \end{matrix} ; \frac{-4x}{(1-x)^2} \right].$$

To prove this result, Andrews and Stanton used two Bailey cubic ${}_3F_2$ transformation formulas, both of which satisfy the $(*)$ condition. Consequently, our dictionary method allows us to translate the proofs to obtain the following result.

THEOREM 9.12. *For a given finite field $\mathbb{F}_q$ of odd characteristic, let $A, B \in \widehat{\mathbb{F}_q^\times}$ such that none of A, B, $A\overline{B}^2$, $A^2\overline{B}^3$, $\phi A\overline{B}$, $\phi A\overline{B}^3$ is trivial and x an element in $\mathbb{F}_q$ with $x \neq \pm 1, \tfrac{1}{2}$. We have*

$$ {}_3\mathbb{F}_2 \left[\begin{matrix} A & B & A\phi\overline{B} \\ & B^2 & A^2\overline{B}^2 \end{matrix} ; 4x(1-x) \right] = \overline{A}^2(1-x) \, {}_3F_2 \left[\begin{matrix} A & B & A\phi\overline{B} \\ & B^2 & A^2\overline{B}^2 \end{matrix} ; \frac{-4x}{(1-x)^2} \right].$$

We note that the conditions in Theorem 9.12 are used in order to avoid the need for delta terms. The result could be stated for a less restricted class of characters, but would not appear as clean. Before proving Theorem 9.12, we require some additional results. We start by using our approach to obtain a finite field analogue of the Bailey cubic transformations.

9.4.1. Finite field analogues of Bailey cubic transformations.
There are two Bailey cubic transforms (see [**10**, (4.06)]):

$$(9.18) \quad (1-x)^{-a} \, {}_3F_2 \left[\begin{matrix} \tfrac{a}{3} & \tfrac{a+1}{3} & \tfrac{a+2}{3} \\ & b & a-b+\tfrac{3}{2} \end{matrix} ; \frac{27x^2}{4(1-x)^3} \right] =$$
$$ {}_3F_2 \left[\begin{matrix} a & b-\tfrac{1}{2} & a+1-b \\ & 2b-1 & 2a+2-2b \end{matrix} ; 4x \right],$$

9.4. A FINITE FIELD ANALOGUE OF A THEOREM OF ANDREWS AND STANTON

and also (see [**10**, (4.05)])

$$(9.19) \quad (1-x)^{-a} \, {}_3F_2\left[\begin{matrix} \frac{a}{3} & \frac{a+1}{3} & \frac{a+2}{3} \\ & a-b+\frac{3}{2} & b \end{matrix}; \frac{-27x}{4(1-x)^3}\right] =$$
$$ {}_3F_2\left[\begin{matrix} a & 2b-a-1 & a+2-2b \\ & b & a+\frac{3}{2}-b \end{matrix}; \frac{x}{4}\right].$$

We note that both (9.18) and (9.19) satisfy the (∗) condition (see Appendix §12.1 for more details).

We first consider the $\mathbb{F}_q$ version of (9.18). To do this, we apply the dictionary from §2.4. We use B for the $\mathbb{F}_q$ analogue of $b - \frac{1}{2}$ and A for the analogue of $a/3$.

Theorem 9.13. *Let $q \equiv 1 \pmod{6}$ be a prime power. Let η_3 be a primitive character of order 3 on $\mathbb{F}_q$, and $A, B \in \widehat{\mathbb{F}_q^\times}$ satisfying A^3, B, $A^3\overline{B}^2$, $A^6\overline{B}^3$, $\phi A^3\overline{B}$, $\phi A^3\overline{B}^3 \neq \varepsilon$. Then we have*

$$\overline{A}^3(1-x) \, {}_3\mathbb{F}_2\left[\begin{matrix} A & A\eta_3 & A\overline{\eta_3} \\ & \phi B & A^3\overline{B} \end{matrix}; \frac{27x^2}{4(1-x)^3}\right] = {}_3\mathbb{F}_2\left[\begin{matrix} A^3 & B & \phi A^3\overline{B} \\ & B^2 & A^6\overline{B}^2 \end{matrix}; 4x\right]$$
$$- \delta(x+2)\frac{g(\phi)g(\overline{A}^3)}{g(\phi\overline{B})g(\overline{A}^3 B)} - \delta(1-x) \sum_{\substack{\chi \in \widehat{\mathbb{F}_q^\times} \\ \chi^3 = \overline{A}^3}} \frac{g(\overline{\chi})g(\phi\overline{B}\chi)g(B\overline{A^3\chi})}{g(\phi\overline{B})g(\overline{A}^3 B)g(A^3)} \overline{\chi}(-4).$$

This is a special case of the following identity

Theorem 9.14. *Let q be an odd prime power and let $A, B \in \widehat{\mathbb{F}_q^\times}$ be such that A, B, $A\overline{B}^2$, $A^2\overline{B}^3$, $\phi A\overline{B}$, $\phi A\overline{B}^3 \neq \varepsilon$. Then we have*

$${}_3\mathbb{F}_2\left[\begin{matrix} A & B & \phi A\overline{B} \\ & B^2 & A^2\overline{B}^2 \end{matrix}; 4x\right]$$
$$= \frac{1}{(q-1)^2} \frac{\overline{A}(1-x)}{g(\phi\overline{B})g(\overline{A}B)} \sum_{\chi \in \widehat{\mathbb{F}_q^\times}} (A)_{\chi^3} \, g(\overline{\chi})g(\phi\overline{B}\chi)g(B\overline{A\chi}) \cdot \chi\left(\frac{x^2}{4(x-1)^3}\right)$$
$$+ \delta(x) + \delta(x+2)\frac{g(\phi)g(\overline{A})}{g(\phi\overline{B})g(\overline{A}B)} + \delta(1-x) \sum_{\substack{\chi \in \widehat{\mathbb{F}_q^\times} \\ \chi^3 = \overline{A}}} \frac{g(\overline{\chi})g(\phi\overline{B}\chi)g(B\overline{A\chi})}{g(\phi\overline{B})g(\overline{A}B)g(A)} \overline{\chi}(-4).$$

As an immediate consequence, one has the following evaluation formula by letting $x = 1$.

Corollary 9.15. *Given the assumptions as in Theorem 9.14,*

$${}_3\mathbb{F}_2\left[\begin{matrix} A & B & \phi A\overline{B} \\ & B^2 & A^2\overline{B}^2 \end{matrix}; 4\right] = \sum_{\substack{\chi \in \widehat{\mathbb{F}_q^\times} \\ \chi^3 = \overline{A}}} \frac{g(\overline{\chi})g(\phi\overline{B}\chi)g(B\overline{A\chi})}{g(\phi\overline{B})g(\overline{A}B)g(A)} \overline{\chi}(-4).$$

The proof of Theorem 9.14 is mainly based on Corollary 9.9, the quadratic version of Pfaff-SaalSchütz formula. Before we begin the proof, we introduce some notation and prove two short lemmas which will assist in the body of the proof. If $A, B, K \in \widehat{\mathbb{F}_q^\times}$, define

(9.20) $$I_{A,B}(K) := \frac{1}{q-1} \sum_{\chi \in \widehat{\mathbb{F}_q^\times}} g(AK\chi)g(\phi\overline{B\chi})g(\overline{AB}\overline{\chi})g(\overline{K}\chi^2)g(\overline{\chi})\overline{\chi}(-4),$$

and

$$J_{A,B}(K) := K(4)J(AK, \overline{K})J(BK, \overline{B^2K})J(\phi A\overline{B}K, B^2\overline{A^2}K).$$

With this notation in place, we state and prove two lemmas. The first evaluates the above character sums for special choices of K.

LEMMA 9.16. *Under the same assumptions on A, B as Theorem 9.14, for any characters K and a fixed finite field $\mathbb{F}_q$, the following identities hold.*

(1) *If $K = \overline{B}$,*

$$I_{A,B}(\overline{B}) = -g(\phi)qA(-1)B(2) - g(\phi)A\overline{B}(4)J(\phi, A\overline{B})J(B^3\overline{A}^2, \phi\overline{B}^2 A)$$
$$J_{A,B}(\overline{B}) = -\overline{B}(4)J(B, A\overline{B})J(\phi\overline{B}^2 A, B^3\overline{A}^2).$$

(2) *If $K = \overline{A}B$, or $\phi\overline{B}$, then*

$$I_{A,B}(K) = g(\phi)qAK(4)A(-1)J(AK, AK^2) - qA(-1)\overline{K}(2)g(\phi)$$
$$\quad - \delta(K)AK(-1)(q-1)g(\phi)J(\phi, AK);$$
$$J_{A,B}(\overline{A}B) = B\overline{A}(4)A(-1)J(B, A\overline{B})J(B^2\overline{A}, B)J(\phi, B\overline{A}),$$
$$J_{A,B}(\phi\overline{B}) = \overline{B}(4)J(\phi A\overline{B}, \phi B)J(\phi, \phi\overline{B})J(\overline{B}^2 A, \phi B^3 \overline{A}).$$

PROOF. Following the special cases of the Corollary 9.9, we can get $I_{A,B}(K)$ immediately. In particular,

$$\frac{I_{A,B}(\overline{B})}{g(\phi)} = -qA(-1)B(2) - A\overline{B}(4)J(\phi, A\overline{B})J(\overline{A}^2 B^3, \phi A\overline{B}^2)$$
$$\quad - \delta(B)q(q-1)AB(-1) - \delta(\phi A\overline{B})A(-1)(q-1)$$
$$= -qA(-1)B(2) - A\overline{B}(4)J(\phi, A\overline{B})J(\overline{A}^2 B^3, \phi A\overline{B}^2),$$

since $B, \phi A\overline{B}$ are non-trivial characters.
When $K = \overline{A}B$,

$$J_{A,B}(K) = B\overline{A}(4)J(B, A\overline{B})J(B^2\overline{A}, A\overline{B}^3)J(\phi, B\overline{A})$$
$$= B\overline{A}(4)A(-1)J(B, A\overline{B})J(B^2\overline{A}, B)J(\phi, B\overline{A});$$

the other two can be computed from the definition of $J_{A,B}(K)$. □

The next lemma relates $I_{A,B}(K)$ and $J_{A,B}(K)$ under special assumptions.

LEMMA 9.17. *Under the same assumptions on A, B as Theorem 9.14, for any characters K of a fixed finite field $\mathbb{F}_q$ one has*

$$I_{A,B}(K) = A(-4)g(A)g(\overline{B})g(\phi B\overline{A}) \cdot J_{A,B}(K) - A(-1)\overline{K}(2)qg(\phi).$$

9.4. A FINITE FIELD ANALOGUE OF A THEOREM OF ANDREWS AND STANTON

Consequently,

$$\frac{1}{q-1} \sum_{K \in \widehat{\mathbb{F}_q^\times}} K(-x) \cdot I_{A,B}(K) = -\delta(x+2) A(-1) q g(\phi)$$

$$+ \frac{A(4)\phi(-1)}{q} g(A) g(\overline{B}) g(\phi B \overline{A}) \,_3\mathbb{P}_2 \begin{bmatrix} A & B & \phi A \overline{B} \\ & B^2 & A^2 \overline{B}^2 \end{bmatrix} ; 4x \end{bmatrix}.$$

PROOF. When $K^2 \neq \overline{B}^2$, $K \neq \overline{A}B$, by letting $A = \overline{K}$, $B = AK$, and $C = \phi \overline{B}$ in Corollary 9.9, one has

$$I_{A,B}(K) = A(-1) B(4) g(\phi) J(BK, A\overline{B}^2) J(\overline{A}B^2, \overline{B^2 K}) J(\overline{K}, A^2 \overline{B}^2 K^2)$$
$$- A(-1) \overline{K}(2) q g(\phi).$$

Using the next equality, obtained from the duplication and reflection formulas of Gauss sums,

$$J(\overline{K}, A^2 \overline{B}^2 K^2) = A\overline{B}K(4) J(A\overline{B}K, A\overline{B}) J(\phi A \overline{B} K, \overline{K}) / J(\phi, A\overline{B}),$$

one can derive

$$\begin{aligned} I_{A,B}(K) =& A(-1) B(4) g(\phi) \cdot AK(-1) J(BK, \overline{AB K}) J(\overline{B^2 K}, AK) \\ & \cdot \frac{A\overline{B}K(4)}{J(\phi, A\overline{B})} J(A\overline{B}K, B^2 \overline{A^2 K}) J(\phi A \overline{B} K, \overline{K}) - A(-1) \overline{K}(2) q g(\phi) \\ =& A(-1) B(4) g(\phi) \left(\frac{AB(-1) A\overline{B}(4)}{J(\phi, A\overline{B})} \frac{g(A) g(\overline{B}) g(\phi B \overline{A})}{g(B\overline{A}) g(\phi A \overline{B})} - \delta(B\overline{A}) A(-1)(q-1) \right) \\ & \cdot J_{A,B}(K) - A(-1) \overline{K}(2) q g(\phi). \end{aligned}$$

The claim follows from straightforward computation. For the other case, the claim follows from the Lemma 9.16. $\square$

We are now ready to prove Theorem 9.14.

PROOF OF THEOREM 9.14. When $x = 0$, both sides of the desired equation have value 1. Assume $x \neq 0$. The character sum

$$\frac{1}{(q-1)^2} \frac{\overline{A}(1-x)}{g(\phi \overline{B}) g(\overline{A}B)} \sum_{\chi \in \widehat{\mathbb{F}_q^\times}} (A)_{\chi^3} \, g(\overline{\chi}) g(\phi \overline{B} \chi) g(B\overline{A}\chi) \chi\left(\frac{x^2}{4(x-1)^3} \right)$$

can be written as

$$\frac{1}{(q-1)^2}\frac{\overline{A}(1-x)}{g(\phi\overline{B})g(\overline{A}B)}\sum_{\chi\in\widehat{\mathbb{F}_q^\times}}(A)_{\chi^3}\,g(\overline{\chi})g(\phi\overline{B\chi}g(B\overline{A\chi})\chi\left(\frac{x^2}{4(x-1)^3}\right)$$

$$=\frac{1}{(q-1)^2}\frac{1}{g(\phi\overline{B})g(\overline{A}B)}$$
$$\cdot\sum_{\chi,\varphi\in\widehat{\mathbb{F}_q^\times}}(A)_{\chi^3}g(\overline{\chi})g(\phi\overline{B\chi}g(B\overline{A\chi})\overline{\chi}(4)\,J(A\chi^3\varphi,\overline{\varphi})\cdot\varphi\chi^2(-x)$$

$$=\frac{1}{(q-1)^2}\frac{1}{g(\phi\overline{B})g(\overline{A}B)}$$
$$\cdot\sum_{\chi,\varphi}(A)_{\chi^3}g(\overline{\chi})g(\phi\overline{B\chi})g(B\overline{A\chi})\overline{\chi}(-4)\,(A\chi^3)_\varphi g(\overline{\varphi})\cdot\varphi\chi^2(-x)$$
$$+\frac{1}{(q-1)}\frac{1}{g(\phi\overline{B})g(\overline{A}B)}$$
$$\cdot\sum_\chi \delta(A\chi^3)(A)_{\chi^3}g(\overline{\chi})g(\phi\overline{B\chi})g(B\overline{A\chi})\overline{\chi}(-4)\,\chi^2(x)\sum_\varphi\varphi(x).$$

By using (2.19), which says $(A)_{\chi_1\chi_2}=(A)_{\chi_1}(A\chi_1)_{\chi_2}$, the above equals

$$\frac{1}{(q-1)^2}\frac{1}{g(\phi\overline{B})g(\overline{A}B)}\sum_{\chi,\varphi\in\widehat{\mathbb{F}_q^\times}}(A)_{\chi^3\varphi}g(\overline{\chi})g(\phi\overline{B\chi})g(B\overline{A\chi})\overline{\chi}(-4)\,g(\overline{\varphi})\cdot\varphi\chi^2(-x)$$
$$-\delta(1-x)\frac{1}{g(\phi\overline{B})g(\overline{A}B)g(A)}\sum_\chi \delta(A\chi^3)g(\overline{\chi})g(\phi\overline{B\chi})g(B\overline{A\chi})\overline{\chi}(-4)$$
$$\stackrel{K:=\varphi\chi^2}{=}\frac{1}{(q-1)^2}\frac{1}{g(\phi\overline{B})g(\overline{A}B)}\sum_{\chi,K\in\widehat{\mathbb{F}_q^\times}}(A)_{\chi K}g(\overline{\chi})g(\phi\overline{B\chi})g(B\overline{A\chi})\overline{\chi}(-4)\,g(\overline{K}\chi^2)\cdot K(-x)$$
$$-\delta(1-x)\sum_{\chi^3=\overline{A}}\frac{g(\overline{\chi})g(\phi\overline{B\chi})g(B\overline{A\chi})}{g(\phi\overline{B})g(\overline{A}B)g(A)}\overline{\chi}(-4).$$

We write the first term above as

$$\frac{1}{(q-1)}\frac{1}{g(\phi\overline{B})g(\overline{A}B)g(A)}\sum_{K\in\widehat{\mathbb{F}_q^\times}}I_{A,B}(K)\cdot K(-x),$$

where $I_{A,B}(K)$ is defined in (9.20). By applying Lemma 9.17, we have

$$\frac{1}{(q-1)^2}\frac{\overline{A}(1-x)}{g(\phi\overline{B})g(\overline{A}B)}\sum_{\chi\in\widehat{\mathbb{F}_q^\times}}(A)_{\chi^3}\,g(\overline{\chi})g(\phi\overline{B\chi}g(B\overline{A\chi})\chi\left(\frac{x^2}{4(x-1)^3}\right)$$
$$={}_3\mathbb{F}_2\left[\begin{matrix}A & B & \phi A\overline{B}\\ & B^2 & A^2\overline{B}^2\end{matrix}\,;\,4x\right]-\delta(x)$$
$$-\delta(x+2)\frac{g(\phi)g(\overline{A})}{g(\phi\overline{B})g(\overline{A}B)}-\delta(1-x)\sum_{\chi^3=\overline{A}}\frac{g(\overline{\chi})g(\phi\overline{B\chi})g(B\overline{A\chi})}{g(\phi\overline{B})g(\overline{A}B)g(A)}\overline{\chi}(-4),$$

9.4. A FINITE FIELD ANALOGUE OF A THEOREM OF ANDREWS AND STANTON

since
$$\frac{A(4)\phi(-1)}{q}\frac{g(\overline{B})g(\phi B\overline{A})}{g(\phi\overline{B})g(\overline{A}B)}J(B,B)J(\phi A\overline{B},\phi A\overline{B})$$
$$=\frac{A(-4)}{q}\frac{g(\overline{B})g(\phi B\overline{A})}{g(\phi\overline{B})g(\overline{A}B)}J(B,\overline{B}^2)J(\phi A\overline{B},\overline{A}^2B^2)$$
$$=1.$$

□

PROOF OF THEOREM 9.13. When $x = 0$, both sides take value 1. Thus, assume $x \neq 0$. From the definition, we derive that
$$\overline{A}^3(1-x)\,_3\mathbb{F}_2\left[\begin{matrix} A & A\eta_3 & A\overline{\eta_3} \\ & \phi B & A^3\overline{B} \end{matrix};\frac{27x^2}{4(1-x)^3}\right]$$
$$=\frac{\overline{A}^3(1-x)}{(q-1)^2J(A\eta_3,\phi\overline{B})J(A\overline{\eta_3},\overline{A}^3B)}$$
$$\sum_{\chi,\varphi\in\widehat{\mathbb{F}_q^\times}}J(A\chi,\overline{\chi})J(A\eta_3\chi,\phi\overline{B}\chi)J(A\overline{\eta_3}\chi,B\overline{A}^3\chi)\chi(27)\chi\left(\frac{x^2}{4(x-1)^3}\right)$$
$$=\frac{1}{(q-1)^2}\frac{\overline{A}^3(1-x)}{g(\phi\overline{B})g(\overline{A}^3B)}$$
$$\cdot\sum_{\chi,\varphi}(A)_\chi(A\eta_3)_\chi(A\overline{\eta_3})_\chi g(\overline{\chi})g(\phi\overline{B}\chi)g(B\overline{A}^3\chi)\chi(27)\chi\left(\frac{x^2}{4(x-1)^3}\right)$$
$$=\frac{1}{(q-1)^2}\frac{\overline{A}^3(1-x)}{g(\phi\overline{B})g(\overline{A}^3B)}\cdot\sum_{\chi,\varphi}(A^3)_{\chi^3}g(\overline{\chi})g(\phi\overline{B}\chi)g(B\overline{A}^3\chi)\overline{\chi}\left(\frac{x^2}{4(x-1)^3}\right).$$

The theorem follows from Theorem 9.14 by replacing A with A^3. □

Theorem 9.14 provides a finite field analogue for (9.18), and we will now use it to obtain a finite field analogue for (9.19).

THEOREM 9.18. *Let q be a power of odd prime, and $A, B \in \widehat{\mathbb{F}_q^\times}$ satisfying A, B, $A\overline{B}^2$, $A^2\overline{B}^3$, $\phi A\overline{B}$, $\phi A\overline{B}^3 \neq \varepsilon$. Then we have*
$$_3\mathbb{F}_2\left[\begin{matrix} A & A\overline{B}^2 & \overline{A}B^2 \\ & A\overline{B} & \phi B \end{matrix};\frac{x}{4}\right]$$
$$=\frac{1}{(q-1)^2}\frac{\overline{A}(1-x)}{g(\phi\overline{B})g(\overline{A}B)}\sum_{\chi\in\widehat{\mathbb{F}_q^\times}}(A)_{\chi^3}\,g(\overline{\chi})g(\phi\overline{B}\chi)g(B\overline{A}\chi)\cdot\chi\left(\frac{x}{4(1-x)^3}\right)$$
$$+\delta(x)-\delta(x+1/2)A(2)\frac{g(\phi)g(\overline{A})}{g(\phi\overline{B})g(\overline{A}B)}-\delta(1-x)\sum_{\substack{\chi\in\widehat{\mathbb{F}_q^\times}\\\chi^3=\overline{A}}}\frac{g(\overline{\chi})g(\phi\overline{B}\chi)g(B\overline{A}\chi)}{g(\phi\overline{B})g(\overline{A}B)g(A)}\overline{\chi}(4).$$

As an immediate consequence, one has the following evaluation formula by letting $x = 1$, comparable to Corollary 9.15.

COROLLARY 9.19. *Given the assumptions as in Theorem 9.18, we have*

$$_3\mathbb{F}_2\begin{bmatrix} A & A\overline{B}^2 & \overline{A}B^2 \\ & A\overline{B} & \phi B \end{bmatrix}; \frac{1}{4}\end{bmatrix} = \sum_{\substack{\chi \in \widehat{\mathbb{F}_q^\times} \\ \chi^3 = \overline{A}}} \frac{g(\overline{\chi})g(\phi\overline{B}\chi)g(B\overline{A}\chi)}{g(\phi\overline{B})g(\overline{A}B)g(A)}\overline{\chi}(4).$$

PROOF OF THEOREM 9.18. If we do a variable change $\chi \mapsto \overline{A}\chi$ in the character sum corresponding to the $_3\mathbb{P}_2$-period function on the left hand side of the identity in Theorem 9.14, we obtain that when $x \neq 0$,

$$(9.21) \qquad _3\mathbb{P}_2\begin{bmatrix} A & B & A\phi\overline{B} \\ & B^2 & A^2\overline{B}^2 \end{bmatrix}; 4x \end{bmatrix} = \phi(-1)\overline{A}(4x) {}_3\mathbb{P}_2\begin{bmatrix} A & A\overline{B}^2 & \overline{A}B^2 \\ & A\overline{B} & \phi B \end{bmatrix}; \frac{1}{4x}\end{bmatrix}.$$

By letting $x \mapsto \frac{1}{x}$ on the right hand sides of Theorem 9.14 and (9.21) we see that ignoring the delta terms gives

$$_3\mathbb{F}_2\begin{bmatrix} A & A\overline{B}^2 & A\overline{B}^2 \\ & A\overline{B} & \phi B \end{bmatrix}; \frac{x}{4}\end{bmatrix}$$

$$= \frac{J(B,B)J(A\phi\overline{B}, A\phi\overline{B})}{J(A\overline{B}^2, B)J(\overline{A}B^2, \phi A\overline{B})} \cdot \phi(-1)A\left(\frac{4}{x}\right)$$

$$\cdot \frac{1}{(q-1)^2} \frac{\overline{A}\left(\frac{x-1}{x}\right)}{g(\phi\overline{B})g(\overline{A}B)} \sum_{\chi \in \widehat{\mathbb{F}_q^\times}} (A)_{\chi^3}\, g(\overline{\chi})g(\phi\overline{B}\chi)g(B\overline{A}\chi) \cdot \chi\left(\frac{x}{4(1-x)^3}\right)$$

$$= \phi A(-1)\overline{A}(4)\phi(-1)A\left(\frac{4}{x}\right)$$

$$\cdot \frac{1}{(q-1)^2} \frac{\overline{A}\left(\frac{x-1}{x}\right)}{g(\phi\overline{B})g(\overline{A}B)} \sum_{\chi \in \widehat{\mathbb{F}_q^\times}} (A)_{\chi^3}\, g(\overline{\chi})g(\phi\overline{B}\chi)g(B\overline{A}\chi) \cdot \chi\left(\frac{x}{4(1-x)^3}\right)$$

$$= \frac{1}{(q-1)^2} \frac{\overline{A}(1-x)}{g(\phi\overline{B})g(\overline{A}B)} \sum_{\chi \in \widehat{\mathbb{F}_q^\times}} (A)_{\chi^3}\, g(\overline{\chi})g(\phi\overline{B}\chi)g(B\overline{A}\chi) \cdot \chi\left(\frac{x}{4(1-x)^3}\right)$$

$\square$

As a corollary of Theorem 9.18, we can obtain a finite field analogue for Equation (9.19).

THEOREM 9.20. *Let $q \equiv 1 \pmod 6$ be a prime power. Let η_3 be a primitive character of order 3 on $\mathbb{F}_q$, and A, B be characters satisfying A^3, B, $A^3\overline{B}^2$, $A^6\overline{B}^3$, $\phi A^3\overline{B}$, $\phi A^3\overline{B}^3 \neq \varepsilon$. Then we have*

$$\overline{A}^3(1-x) {}_3\mathbb{F}_2\begin{bmatrix} A & A\eta_3 & A\overline{\eta_3} \\ & \phi B & A^3\overline{B} \end{bmatrix}; \frac{-27x}{4(1-x)^3}\end{bmatrix} = {}_3\mathbb{F}_2\begin{bmatrix} A^3 & A^3\overline{B}^2 & \overline{A}^3 B^2 \\ & A^3\overline{B} & \phi B \end{bmatrix}; \frac{x}{4}\end{bmatrix}$$

$$- \delta(x+1/2)A^3(2)\frac{g(\phi)g(\overline{A}^3)}{g(\phi\overline{B})g(\overline{A}^3B)} - \delta(1-x) \sum_{\substack{\chi \in \widehat{\mathbb{F}_q^\times} \\ \chi^3 = \overline{A}^3}} \frac{g(\overline{\chi})g(\phi\overline{B}\chi)g(B\overline{A^3}\chi)}{g(\phi\overline{B})g(\overline{A}^3B)g(A^3)}\overline{\chi}(4).$$

We are now in the position to prove Theorem 9.12, the finite field analogue of Theorem 9.11. We follow the ideas given by Andrews and Stanton in [**7**].

PROOF OF THEOREM 9.12. Following the ideas of the proof of Theorem 9.11 by Andrews and Stanton [7], let $x = y^2/(y-1)$ in Theorem 9.18 and $x = y(1-y)$ in Theorem 9.14. Notice that the function

$$F(y) := {}_3\mathbb{F}_2 \begin{bmatrix} A & A\overline{B}^2 & \overline{A}B^2 \\ & A\overline{B} & \phi B \end{bmatrix}; \frac{y^2}{4(y-1)} \end{bmatrix}$$

$$= \frac{1}{(q-1)^2} \frac{\overline{A}\left(1 - \frac{y^2}{y-1}\right)}{g(\phi\overline{B})g(\overline{A}B)} \sum_{\chi \in \widehat{\mathbb{F}_q^\times}} (A)_{\chi^3} g(\overline{\chi}) g(\phi \overline{B}\chi) g(B\overline{A}\chi) \cdot \chi\left(\frac{y^2(y-1)^2}{4(y-1-y^2)^3}\right)$$

is equal to

$$A(1-y)\, {}_3\mathbb{F}_2 \begin{bmatrix} A & B & A\phi\overline{B} \\ & B^2 & A^2\overline{B}^2 \end{bmatrix}; 4y(1-y) \end{bmatrix},$$

if $y \neq 0, \pm 1, \frac{1}{2}$. As the rational function $\frac{y^2}{4(y-1)}$ is invariant under the change of variable $y \mapsto y/(y-1)$, i.e. $F(y) = F(y/(y-1))$, we have

$${}_3\mathbb{F}_2 \begin{bmatrix} A & B & A\phi\overline{B} \\ & B^2 & A^2\overline{B}^2 \end{bmatrix}; 4y(1-y) \end{bmatrix} = \overline{A}((1-y)^2)\, {}_3\mathbb{F}_2 \begin{bmatrix} A & B & A\phi\overline{B} \\ & B^2 & A^2\overline{B}^2 \end{bmatrix}; \frac{-4y}{(1-y)^2} \end{bmatrix},$$

for any element $y \neq 0, \pm 1, \frac{1}{2}$. □

9.5. Another application of Bailey cubic transformations

In this section, we consider a degenerate case of (9.19) corresponding to the triangle group (see (3.11)) $(2,3,4)$ and prove a finite field version of this identity. We close the section by providing conjectures, based on numerical evidence, for finite field versions of identities related to the triangle group $(2,3,3)$.

Letting $b = \frac{a+2}{3}$ in (9.19) gives the following degenerate case of the Bailey cubic transformation formula:

$$(9.22) \qquad {}_2F_1 \begin{bmatrix} a & \frac{1-a}{3} \\ & \frac{4a+5}{6} \end{bmatrix}; x \end{bmatrix} = (1-4x)^{-a}\, {}_2F_1 \begin{bmatrix} \frac{a}{3} & \frac{a+1}{3} \\ & \frac{4a+5}{6} \end{bmatrix}; \frac{-27x}{(1-4x)^3} \end{bmatrix},$$

which we note satisfies condition (∗) as it is a degenerate (from ${}_3F_2$ to ${}_2F_1$) case of (9.19) (see Appendix §12.1). Further specifying $a = -\frac{1}{8}$ in (9.22), gives

$${}_2F_1 \begin{bmatrix} -\frac{1}{24} & \frac{7}{24} \\ & \frac{3}{4} \end{bmatrix}; \frac{-27x}{(1-4x)^3} \end{bmatrix} = (1-4x)^{-\frac{1}{8}}\, {}_2F_1 \begin{bmatrix} -\frac{1}{8} & \frac{3}{8} \\ & \frac{3}{4} \end{bmatrix}; x \end{bmatrix},$$

and then using (1.3) with $a = -\frac{3}{8}$ to evaluate the right hand side gives the following algebraic identity corresponding to the triangle group $(2,3,4)$:

$$(9.23) \qquad {}_2F_1 \begin{bmatrix} -\frac{1}{24} & \frac{7}{24} \\ & \frac{3}{4} \end{bmatrix}; \frac{-27x}{(1-4x)^3} \end{bmatrix} = \left(\frac{(1+\sqrt{1-x})^2}{4(1-4x)}\right)^{\frac{1}{8}}.$$

Our goal here is to derive the finite field analogue of (9.23).

Recall that in §9.2.1 we outlined how our approach can be used to prove finite field versions of classical formulas satisfying the (∗) condition. We first apply these ideas to obtain the following finite field version of (9.22). Here, the dictionary in §2.4 translates $\frac{a}{3}$ to E. By taking $B = \phi A \overline{\eta}_3$ in Theorem 9.20, we obtain the following:

COROLLARY 9.21. *Let $q \equiv 1 \pmod{3}$ be a prime power, $E \in \widehat{\mathbb{F}_q^\times}$ such that $E^6 \neq \varepsilon$. Let η_3 be a primitive cubic character. Then for $x \neq 0, 1, \frac{1}{4}, -\frac{1}{8}$,*

$$_2F_1\begin{bmatrix} E^3 & \eta_3\overline{E} \\ & E^2\phi\eta_3 \end{bmatrix}; x\end{bmatrix} = \overline{E^3}(1-4x)\,_2F_1\begin{bmatrix} E & E\eta_3 \\ & E^2\phi\eta_3 \end{bmatrix}; \frac{-27x}{(1-4x)^3}\end{bmatrix}.$$

We now establish the finite field analogue of (9.23) using an idea that is parallel to the proof of (9.23) described above.

THEOREM 9.22. *Let $q \equiv 1 \pmod{24}$ be a prime power, and let $x \neq -\frac{1}{8}, \frac{1}{4}$ in $\mathbb{F}_q$. Let η be an order 24 character on $\mathbb{F}_q^\times$. Then*

$$_2F_1\begin{bmatrix} \overline{\eta} & \eta^7 \\ & \eta^{18} \end{bmatrix}; \frac{-27x}{(1-4x)^3}\end{bmatrix} =$$
$$\left(\frac{1+\phi(1-x)}{2}\right)\left(\eta^3\left(\frac{(1+\sqrt{1-x})^2}{2^2(1-4x)}\right) + \eta^3\left(\frac{(1-\sqrt{1-x})^2}{2^2(1-4x)}\right)\right).$$

PROOF. It is easy to see that when $x = 0$, both sides equal to 1.

When $x = 1$, using Gauss evaluation formula and [**14**, Theorem 3.6.6], one has

$$_2F_1\begin{bmatrix} \eta^7 & \overline{\eta} \\ & \eta^{18} \end{bmatrix}; 1\end{bmatrix} = \frac{J(\phi,\overline{\eta})}{J(\overline{\eta},\overline{\eta}^5)} = \eta^3\left(-\frac{4}{3}\right) = \overline{\eta}^3\left(-3\cdot 2^2\right),$$

since 2 is a square in $\mathbb{F}_q$. Hence the claim holds when $x = 1$.

For $x \neq 0, 1, \frac{1}{4}, -\frac{1}{8}$, by Corollary 9.21, we have

$$_2F_1\begin{bmatrix} \overline{\eta} & \eta^7 \\ & \eta^{18} \end{bmatrix}; \frac{-27x}{(1-4x)^3}\end{bmatrix} = \overline{\eta}^3(1-4x)\,_2F_1\begin{bmatrix} \overline{\eta}^3 & \eta^9 \\ & \eta^{18} \end{bmatrix}; x\end{bmatrix}.$$

By Theorem 8.11,

$$_2F_1\begin{bmatrix} \overline{\eta}^3 & \eta^9 \\ & \eta^{18} \end{bmatrix}; x\end{bmatrix} = \left(\frac{1+\phi(1-x)}{2}\right)\left(\eta^6\left(\frac{1+\sqrt{1-x}}{2}\right) + \eta^6\left(\frac{1-\sqrt{1-x}}{2}\right)\right).$$

Composing this with the proceeding cubic formula, we have

$$_2F_1\begin{bmatrix} \overline{\eta} & \eta^7 \\ & \eta^{18} \end{bmatrix}; \frac{-27x}{(1-4x)^3}\end{bmatrix} = \overline{\eta}^3(1-4x)\,_2F_1\begin{bmatrix} \overline{\eta}^3 & \eta^9 \\ & \eta^{18} \end{bmatrix}; x\end{bmatrix}$$
$$= \left(\frac{1+\phi(1-x)}{2}\right)\overline{\eta}^3(1-4x)\left(\eta^6\left(\frac{1+\sqrt{1-x}}{2}\right) + \eta^6\left(\frac{1-\sqrt{1-x}}{2}\right)\right)$$
$$= \left(\frac{1+\phi(1-x)}{2}\right)\left(\eta^3\left(\frac{(1+\sqrt{1-x})^2}{2^2(1-4x)}\right) + \eta^3\left(\frac{(1-\sqrt{1-x})^2}{2^2(1-4x)}\right)\right),$$

which concludes the proof. □

REMARK 9.23. We obtain another algebraic hypergeometric identity corresponding to the triangle group $(2,3,3)$ from the following transformation

$$_2F_1\begin{bmatrix} \frac{3a}{2} & \frac{3a-1}{2} \\ & a+\frac{1}{2} \end{bmatrix}; -\frac{z^2}{3}\end{bmatrix} = (1+z)^{1-3a}\,_2F_1\begin{bmatrix} a-\frac{1}{3} & a \\ & 2a \end{bmatrix}; \frac{2z(3+z^2)}{(1+z)^3}\end{bmatrix}$$

[**33**, pp. 114 (39)] (or see [**40**]) with $a = \frac{1}{4}$, namely,

$$(9.24) \qquad _2F_1\begin{bmatrix} -\frac{1}{12} & \frac{1}{4} \\ & \frac{1}{2} \end{bmatrix}; \frac{2z(3+z^2)}{(1+z)^3}\end{bmatrix} = \left(\frac{1+\sqrt{1+z^2/3}}{2(1+z)}\right)^{1/4}.$$

Numerical evidence suggests that the finite field analogue of (9.24) is the following.

CONJECTURE 9.24. *Let $q \equiv 1 \pmod{12}$ and η be any order 12 character. Then for $z \in \mathbb{F}_q$ such that $z \neq 0, \pm 1$, and $z^2 + 3 \neq 0$,*

$$_2F_1 \begin{bmatrix} \overline{\eta} & \eta^3 \\ & \phi \end{bmatrix} ; \frac{2z(3+z^2)}{(1+z)^3} \end{bmatrix} \stackrel{?}{=}$$
$$\left(\frac{1 + \phi(1+z^2/3)}{2} \right) \left(\eta^3 \left(\frac{1 + \sqrt{1+z^2/3}}{2(1+z)} \right) + \eta^3 \left(\frac{1 - \sqrt{1+z^2/3}}{2(1+z)} \right) \right).$$

Related to the triangle group $(2, 3, 3)$, Vidūnas gives the following formula, see [**98**, pp. 165],

$$(9.25) \qquad _2F_1 \begin{bmatrix} \frac{1}{4} & -\frac{1}{12} \\ & \frac{2}{3} \end{bmatrix} ; \frac{x(x+4)^3}{4(2x-1)^3} \end{bmatrix} = (1-2x)^{-1/4}.$$

We observe numerically the following $\mathbb{F}_q$ analogue:

CONJECTURE 9.25. *For $q \equiv 1 \pmod{12}$ and η any primitive order 12 character:*

$$_2F_1 \begin{bmatrix} \overline{\eta} & \eta^3 \\ & \overline{\eta}^4 \end{bmatrix} ; \left(\frac{4u(u^3+1)}{8u^3-1} \right)^3 \end{bmatrix} \stackrel{?}{=} 2 \cdot \overline{\eta}^3 (1 - 8u^3),$$

if $\left(4u(u^3+1)/(8u^3-1)\right)^3 \neq 0, 1$, and $8u^3 \neq 1$ in $\mathbb{F}_q$.

Here, we replace x by $4u^3$ in (9.25) so that the pattern for the $\mathbb{F}_q$ version is cleaner. In [**98**], there are several other algebraic hypergeometric identities. It will be nice to obtain their $\mathbb{F}_q$ analogues. We leave these to the interested reader.

9.6. Another cubic $_2F_1$ formula and a corollary

In this section, we will use a different approach to derive a finite field analogue of the following cubic formula by Gessel and Stanton [**38**, (5.18)],

$$(9.26) \qquad _2F_1 \begin{bmatrix} a & -a \\ & \frac{1}{2} \end{bmatrix} ; \frac{27x(1-x)^2}{4} \end{bmatrix} = {_2F_1} \begin{bmatrix} 3a & -3a \\ & \frac{1}{2} \end{bmatrix} ; \frac{3x}{4} \end{bmatrix},$$

which satisfies the $(*)$ condition. The finite field analogue of this formula is stated as Theorem 1.2 in the introduction. Then we will obtain a consequential $_3F_2$ evaluation formula, namely Theorem 1.3, in §9.6.2.

9.6.1. Another cubic $_2F_1$ formula. A proof of (9.26) illustrating the $(*)$ condition is discussed in Appendix §12.2. Though our dictionary method is applicable, we choose to give an alternative proof below, which is applicable to all characters A.

THEOREM 9.26. *Let q be an odd prime power, and $A \in \widehat{\mathbb{F}_q^\times}$. Then for all $x \in \mathbb{F}_q$,*

$$_2F_1 \begin{bmatrix} A & \overline{A} \\ & \phi \end{bmatrix} ; \frac{27x(1-x)^2}{4} \end{bmatrix} =$$
$$_2F_1 \begin{bmatrix} A^3 & \overline{A}^3 \\ & \phi \end{bmatrix} ; \frac{3x}{4} \end{bmatrix} - \phi(-3)\delta(x-1) - \phi(-3)A(-1)\delta(x-4/3).$$

PROOF. It is straightforward to check that the formula holds when $x = 0, 1, \frac{4}{3}$. We thus assume $x \neq 0, 1, \frac{4}{3}$. Let $\lambda := \frac{27x(1-x)^2}{4}$, and $z := \frac{\lambda}{\lambda-1} = \frac{-3x}{4-3x}\left(\frac{3x-3}{3x-1}\right)^2$, so that $1 - z = \frac{1}{1-\lambda}$. For any multiplicative character A, we obtain from Proposition 8.5 that

$$(9.27) \quad {}_2\mathbb{F}_1\begin{bmatrix} A & \overline{A} \\ & \phi \end{bmatrix}; \frac{27x(1-x)^2}{4}\end{bmatrix}$$
$$= \overline{A}(1-\lambda)\,{}_2\mathbb{F}_1\begin{bmatrix} A & \phi A \\ & \phi \end{bmatrix}; z\end{bmatrix} + \delta(1-\lambda)\frac{J(\phi, \overline{A})}{J(\overline{A}, \phi A)}.$$

Furthermore, if $A^2 \neq \varepsilon$, we have

$$ {}_2\mathbb{F}_1\begin{bmatrix} A & \phi A \\ & \phi \end{bmatrix}; z\end{bmatrix} = \left(\frac{1+\phi(z)}{2}\right)\left(\overline{A}^2(1+\sqrt{z}) + \overline{A}^2(1-\sqrt{z})\right),$$

which is given by Theorem 8.11.

If z is not a square, i.e. $\phi(z) = -1$, then the value ${}_2\mathbb{F}_1\begin{bmatrix} A & \phi A \\ & \phi \end{bmatrix}; z\end{bmatrix}$ is 0. When $\phi(z) = 1$, we write $\sqrt{z} = t\left(\frac{t^2+3}{1+3t^2}\right)$ with $T := t^2 = \frac{-3x}{4-3x}$. Then

$$1 + \sqrt{z} = \frac{(1+t)^3}{3t^2+1}, \quad 1 - \sqrt{z} = \frac{(1-t)^3}{3t^2+1},$$

and thus

$$(9.28) \quad \overline{A}\left(\frac{1}{1-z}\right){}_2\mathbb{F}_1\begin{bmatrix} A & \phi A \\ & \phi \end{bmatrix}; z\end{bmatrix} = \overline{A}^3\left(\frac{1}{1-t^2}\right)\left(\overline{A}^6(1+t) + \overline{A}^6(1-t)\right)$$
$$= \overline{A}^3\left(\frac{1}{1-T}\right)\left(\overline{A}^6(1+\sqrt{T}) + \overline{A}^6(1-\sqrt{T})\right).$$

Note that if A is a primitive character of order 6 and $\phi(z) = \pm 1$, then

$${}_2\mathbb{F}_1\begin{bmatrix} A & \overline{A} \\ & \phi \end{bmatrix}; \frac{27x(1-x)^2}{4}\end{bmatrix} = \left(\frac{1+\phi(z)}{2}\right)\cdot\phi\left(\frac{1}{1-T}\right)\left(\varepsilon(1+\sqrt{T}) + \varepsilon(1-\sqrt{T})\right)$$
$$= \left(\frac{1+\phi(z)}{2}\right)\cdot\phi(4-3x)\,2,$$

since T won't be 1. Moreover, when $\phi(z) \neq 0$, we can write

$$\phi(z) = \phi\left(\frac{-3x}{4-3x}\left(\frac{3x-3}{3x-1}\right)^2\right) = \phi(-3x)\phi(4-3x).$$

Therefore,

$${}_2\mathbb{F}_1\begin{bmatrix} A & \overline{A} \\ & \phi \end{bmatrix}; \frac{27x(1-x)^2}{4}\end{bmatrix} = (1 + \phi(-3x)\phi(4-3x))\,\phi(4-3x)$$
$$= \phi(-3x) + \phi(4-3x),$$

when $\phi(z) \neq 0$.

9.6. ANOTHER CUBIC $_2\mathbb{F}_1$ FORMULA AND A COROLLARY

According to Proposition 8.6, for each x, one has

$$_2\mathbb{F}_1\begin{bmatrix} A^3 & \overline{A}^3 \\ & \phi \end{bmatrix}; \frac{3x}{4}\end{bmatrix} = {_2\mathbb{F}_1}\begin{bmatrix} \phi & \phi \\ & \phi \end{bmatrix}; \frac{3x}{4}\end{bmatrix} = \phi(4-3x) + \phi(-3x),$$

and thus we can obtain

$$_2\mathbb{F}_1\begin{bmatrix} A & \overline{A} \\ & \phi \end{bmatrix}; \frac{27x(1-x)^2}{4}\end{bmatrix} =$$

$$_2\mathbb{F}_1\begin{bmatrix} A^3 & \overline{A}^3 \\ & \phi \end{bmatrix}; \frac{3x}{4}\end{bmatrix} - \phi(-3)\delta(x-1) - \phi(3)\delta(x-1/3).$$

Similarly, if A is a primitive character of order 3 and $\phi(z) = \pm 1$, then

$$_2\mathbb{F}_1\begin{bmatrix} A & \overline{A} \\ & \phi \end{bmatrix}; \frac{27x(1-x)^2}{4}\end{bmatrix} = \left(\frac{1+\phi(z)}{2}\right) \cdot 2\varepsilon(3x-4)$$
$$= (1 + \phi(3x)\phi(3x-4))(1 - \delta(3x-4))$$
$$= 1 + \phi(3x)\phi(3x-4) - \delta(3x-4),$$

and

$$_2\mathbb{F}_1\begin{bmatrix} A^3 & \overline{A}^3 \\ & \phi \end{bmatrix}; \frac{3x}{4}\end{bmatrix} = {_2\mathbb{F}_1}\begin{bmatrix} \varepsilon & \varepsilon \\ & \phi \end{bmatrix}; \frac{3x}{4}\end{bmatrix} = 1 + \phi(3x)\phi(3x-4).$$

This leads to the identity in the theorem.

For the case $A^2 = \varepsilon$, using the formulas of imprimitive $_2\mathbb{F}_1$-functions in Proposition 8.6, we can derive the desired identities. For example, when $A = \varepsilon$,

$$_2\mathbb{F}_1\begin{bmatrix} A & \overline{A} \\ & \phi \end{bmatrix}; \frac{27x(1-x)^2}{4}\end{bmatrix} = 1 - \phi(3x)\phi(3x-4)\varepsilon((1-x)(3x-1))$$
$$= 1 - \phi(3x)\phi(3x-4) + \delta(1-x)\phi(-3) + \delta(3x-1)\phi(-3).$$

For the case $A^6 \neq \varepsilon$, we first observe that $\phi(z) = 1$ (resp. -1) if and only if $\phi(T) = 1$ (resp. -1). Hence, if $\phi(z) \neq 0$ (equivalently, $\phi(T) \neq 0$), we have from equation (9.28) and Theorem 8.11 that

$$\overline{A}\left(\frac{1}{1-z}\right) {_2\mathbb{F}_1}\begin{bmatrix} A & \phi A \\ & \phi \end{bmatrix}; z\end{bmatrix}$$
$$= \left(\frac{1+\phi(T)}{2}\right) \overline{A}^3 \left(\frac{1}{1-T}\right) \left(\overline{A}^6(1+\sqrt{T}) + \overline{A}^6(1-\sqrt{T})\right)$$
$$= A^3(1-T) {_2\mathbb{F}_1}\begin{bmatrix} A^3 & \phi A^3 \\ & \phi \end{bmatrix}; T\end{bmatrix}.$$

Thus, for all x, we can derive that

$$\overline{A}\left(\frac{1}{1-z}\right) {}_2\mathbb{F}_1\begin{bmatrix} A & \phi A \\ & \phi \end{bmatrix}; z\end{bmatrix} = A^3(1-T) {}_2\mathbb{F}_1\begin{bmatrix} A^3 & \phi A^3 \\ & \phi \end{bmatrix}; T\end{bmatrix}$$
$$+ \delta(x-1)\left(1 - \left(\frac{1+\phi(T)}{2}\right)\overline{A}^3\left(\frac{1}{1-T}\right)\left(\overline{A}^6(1+\sqrt{T}) + \overline{A}^6(1-\sqrt{T})\right)\right)$$
$$+ \delta(3x-1)\left(0 - \left(\frac{1+\phi(T)}{2}\right)\overline{A}^3\left(\frac{1}{1-T}\right)\left(\overline{A}^6(1+\sqrt{T}) + \overline{A}^6(1-\sqrt{T})\right)\right)$$
$$= A^3(1-T) {}_2\mathbb{F}_1\begin{bmatrix} A^3 & \phi A^3 \\ & \phi \end{bmatrix}; T\end{bmatrix} - \delta(x-1)\phi(-3) - \delta(3x-1)(1+\phi(-3))A(-1),$$

since $(1 \pm \sqrt{-3})^6 = 64$, and $(1 \pm \sqrt{-1/3})^6 = -64/27$.

Using Proposition 8.5, we observe that

$${}_2\mathbb{F}_1\begin{bmatrix} A & \overline{A} \\ & \phi \end{bmatrix}; \frac{27x(1-x)^2}{4}\end{bmatrix}$$
$$= \delta((3x-1)(3x-4))A(-1) - \delta(x-1)\phi(-3) - \delta(3x-1)(1+\phi(-3))A(-1)$$
$$+ A^3(1-T)\left(\overline{A}^3(1-T) {}_2\mathbb{F}_1\begin{bmatrix} A^3 & \overline{A}^3 \\ & \phi \end{bmatrix}; \frac{T}{T-1}\end{bmatrix} + \delta(1-T) \cdot {}_2\mathbb{F}_1\begin{bmatrix} A^3 & \phi A^3 \\ & \phi \end{bmatrix}; 1\end{bmatrix}\right)$$
$$= \delta(3x-4)A(-1) - \delta(x-1)\phi(-3) - \delta(3x-1)\phi(-3)A(-1)$$
$$+ \varepsilon(1-T) {}_2\mathbb{F}_1\begin{bmatrix} A^3 & \overline{A}^3 \\ & \phi \end{bmatrix}; \frac{3x}{4}\end{bmatrix}$$
$$= {}_2\mathbb{F}_1\begin{bmatrix} A^3 & \overline{A}^3 \\ & \phi \end{bmatrix}; \frac{3x}{4}\end{bmatrix} - \delta(x-1)\phi(-3) - \delta(3x-1)\phi(-3)A(-1).$$

$\square$

9.6.2. An application to an evaluation formula. In the next few pages we will obtain a finite field analogue of an evaluation formula by Gessel and Stanton. Our main point for the discussion is as follows. In addition to what can be done for the cases satisfying (∗), there are classical results obtained by methods that have no direct translations to finite fields. However, sometimes the Galois perspective allows us to make predictions which can be verified numerically. To prove them, we need different methods that might appear to be ad hoc when compared with the more systematic approaches we have used so far.

As a corollary of (9.26), Gessel and Stanton showed that for $n \in \mathbb{N}$, $a \in \mathbb{C}$

$$(9.29) \quad {}_3F_2\begin{bmatrix} 1+3a & 1-3a & -n \\ & \frac{3}{2} & -1-3n \end{bmatrix}; \frac{3}{4}\end{bmatrix} = \frac{(1+a)_n(1-a)_n}{(\frac{2}{3})_n(\frac{4}{3})_n}$$
$$= \frac{B(1+a+n, \frac{2}{3})B(1-a+n, \frac{4}{3})}{B(1+a, n+\frac{2}{3})B(1-a, n+\frac{4}{3})}.$$

This result follows from applying Theorem 1 of [**38**] to (9.26) with an additional factor of $x^{-n-2}(1-x)^{-2n-3}(1-3x)$, instead of the factor $x^{-n-2}(1-x)^{-3n-2}(1-3x)$ which is mentioned on page 305 of [**38**]. Although Greene gave a Lagrange inversion formula over finite fields (see Theorem 2.10 for a special case), he pointed out it cannot be used to determine coefficients when the change of variable function is

9.6. ANOTHER CUBIC $_2F_1$ FORMULA AND A COROLLARY

not one-to-one. This means we cannot use Theorem 2.10 to obtain a finite field analogue of (9.29) directly.

Thus to look for a finite field analogue of (9.29), we instead observe that the corresponding Galois representations are expected to be 3-dimensional. Also, if $3a$ corresponds to a character A^3, then there are 3 candidates for the character analogue of a, these being A, $A\eta_3$, and $A\eta_3^2$. This was the language we used when we initially phrased the statement of Theorem 1.3 and then verified numerically.

We now give a proof of Theorem 1.3, which states that if q is a prime power with $q \equiv 1 \pmod{6}$, and $A, \chi, \eta_3 \in \widehat{\mathbb{F}_q^\times}$ such that η_3 has order 3, and none of $A^6, \chi^6, (A\chi)^3, (\overline{A}\chi)^3$ are the trivial character, then

$$_3\mathbb{F}_2\left[\begin{matrix} A^3 & \overline{A}^3 & \chi \\ & \phi & \overline{\chi}^3 \end{matrix}; \frac{3}{4}\right] = \sum_{\substack{B \in \widehat{\mathbb{F}_q^\times} \\ B^3 = A^3}} \frac{J(B\chi, \eta_3)J(\overline{B}\chi, \overline{\eta_3})}{J(B, \chi\eta_3)J(\overline{B}, \chi\overline{\eta_3})} = A(-1)\sum_{B^3 = A^3} \frac{J(B\chi, \overline{B}\chi)}{J(\eta_3\chi, \chi\overline{\eta}_3)}.$$

PROOF OF THEOREM 1.3. Let $\eta := \eta_3$ and $\mathbb{F} := \mathbb{F}_q$ for convenience. Multiplying $\overline{\chi}(x(1-x)^2)$ by the right hand side of Theorem 9.26 and taking the sum over all x, we see

$$\sum_{x \in \mathbb{F}} \overline{\chi}(x(1-x)^2) \, _2F_1\left[\begin{matrix} A^3 & \overline{A}^3 \\ & \phi \end{matrix}; \frac{3x}{4}\right] - \overline{\chi}\left(\frac{4}{27}\right)\phi(-3)A(-1)$$

$$= \frac{1}{J(\overline{A}^3, \phi A^3)} \, _3\mathbb{P}_2\left[\begin{matrix} A^3 & \overline{A}^3 & \chi \\ & \phi & \overline{\chi}^3 \end{matrix}; \frac{3}{4}\right] - \overline{\chi}\left(\frac{4}{27}\right)\phi(-3)A(-1).$$

We now let

$$I := {}_3\mathbb{F}_2\left[\begin{matrix} A^3 & \overline{A}^3 & \chi \\ & \phi & \overline{\chi}^3 \end{matrix}; \frac{3}{4}\right] - \frac{\overline{\chi}(\frac{4}{27})\phi(-3)A(-1)}{J(\overline{\chi}, \overline{\chi}^2)}$$

$$= \sum_{x \in \mathbb{F}} \frac{\overline{\chi}(x(1-x)^2)}{J(\overline{\chi}, \overline{\chi}^2)} \, _2F_1\left[\begin{matrix} A^3 & \overline{A}^3 \\ & \phi \end{matrix}; \frac{3x}{4}\right] - \frac{\overline{\chi}(\frac{4}{27})\phi(-3)A(-1)}{J(\overline{\chi}, \overline{\chi}^2)}.$$

By Theorem 9.26,

$$I = \sum_{x \in \mathbb{F}} \frac{\overline{\chi}(x(1-x)^2)}{J(\overline{\chi}, \overline{\chi}^2)} \, _2F_1\left[\begin{matrix} A & \overline{A} \\ & \phi \end{matrix}; \frac{27x(1-x)^2}{4}\right]$$

$$= \sum_{x \in \mathbb{F}} \frac{\overline{\chi}(x(1-x)^2)}{J(\overline{A}, \phi A)J(\overline{\chi}, \overline{\chi}^2)} \, _2\mathbb{P}_1\left[\begin{matrix} A & \overline{A} \\ & \phi \end{matrix}; \frac{27x(1-x)^2}{4}\right]$$

$$= \sum_{x \in \mathbb{F}} \frac{\overline{\chi}(x(1-x)^2)}{J(\overline{A}, \phi A)J(\overline{\chi}, \overline{\chi}^2)} \frac{A(-1)}{q-1} \sum_{\varphi \in \widehat{\mathbb{F}_q^\times}} J(A\varphi, \overline{\varphi})J(\overline{A}\varphi, \phi\overline{\varphi})\varphi\left(\frac{27}{4}\right)\varphi(x(1-x)^2)$$

$$= \frac{A(-1)}{(q-1)J(\overline{A}, \phi A)J(\overline{\chi}, \overline{\chi}^2)} \sum_{\varphi \in \widehat{\mathbb{F}_q^\times}} J(A\varphi, \overline{\varphi})J(\overline{A}\varphi, \phi\overline{\varphi})\varphi\left(\frac{27}{4}\right) J(\overline{\chi}\varphi, \overline{\chi}^2\varphi^2).$$

9. QUADRATIC OR HIGHER TRANSFORMATION FORMULAS

Using Theorem 2.7, we observe that when $\chi^6 \neq \varepsilon$,

$$\frac{J(\overline{\chi}\varphi, \overline{\chi}^2\varphi^2)}{J(\overline{\chi}, \overline{\chi}^2)} = \frac{g(\overline{\chi}^3)}{g(\overline{\chi})g(\overline{\chi}^2)} \frac{g(\varphi\overline{\chi})g(\varphi^2\overline{\chi}^2)}{g(\varphi^3\overline{\chi}^3)} + \delta\left(\varphi^3\overline{\chi}^3\right) \frac{q-1}{J(\overline{\chi}, \overline{\chi}^2)}$$

$$= \frac{g(\eta\overline{\chi})g(\overline{\eta}\overline{\chi})}{g(\phi\overline{\chi})g(\overline{\chi})} \frac{g(\phi\varphi\overline{\chi})g(\varphi\overline{\chi})}{g(\varphi\overline{\chi}\eta)g(\varphi\overline{\chi}\overline{\eta})} \varphi\left(\frac{4}{27}\right) + \delta\left(\varphi^3\overline{\chi}^3\right) \frac{q-1}{q} \frac{g(\eta\overline{\chi})g(\overline{\eta}\overline{\chi})g(\phi)}{g(\phi\overline{\chi})g(\overline{\chi})} \overline{\chi}\left(\frac{27}{4}\right).$$

Thus the sum I can be divided into two parts, I_1 and I_2, where

$$I_1 = \frac{1}{J(\overline{A}, \phi A)} \frac{A(-1)}{q} \sum_{\substack{\varphi \in \mathbb{F}_q^\times \\ \varphi^3 = \chi^3}} J(A\varphi, \overline{\varphi}) J(\overline{A}\varphi, \phi\overline{\varphi}) \frac{g(\eta\overline{\chi})g(\overline{\eta}\overline{\chi})g(\phi)}{g(\phi\overline{\chi})g(\overline{\chi})} \overline{\chi}\varphi\left(\frac{27}{4}\right)$$

and

$$I_2 = \frac{1}{J(\overline{A}, \phi A)} \frac{A(-1)}{q-1} \sum_{\varphi \in \widehat{\mathbb{F}_q^\times}} J(A\varphi, \overline{\varphi}) J(\overline{A}\varphi, \phi\overline{\varphi}) \frac{g(\eta\overline{\chi})g(\overline{\eta}\overline{\chi})}{g(\phi\overline{\chi})g(\overline{\chi})} \frac{g(\phi\varphi\overline{\chi})g(\varphi\overline{\chi})}{g(\varphi\overline{\chi}\eta)g(\varphi\overline{\chi}\overline{\eta})}.$$

Then under our assumptions, we have

$$I_1 = \frac{A(-1)}{q^2} \sum_{\varphi^3 = \chi^3} g(\eta\overline{\chi})g(\overline{\eta}\overline{\chi}) g(A\varphi)g(\overline{A}\varphi) \frac{g(\overline{\varphi})g(\phi\overline{\varphi})}{g(\phi\overline{\chi})g(\overline{\chi})} \overline{\chi}\varphi\left(\frac{27}{4}\right)$$

$$= \frac{A(-1)}{q} \sum_{\varphi^3 = \chi^3} J(A\varphi, \overline{A}\varphi) J(\eta\overline{\chi}, \overline{\eta}\overline{\chi}) \overline{\chi}\varphi\,(27) = A(-1) \sum_{\varphi^3 = \chi^3} \frac{J(A\varphi, \overline{A}\varphi)}{J(\overline{\eta}\chi, \eta\chi)}.$$

For the sum I_2, we will divide it into 3 parts. First, we observe that

$$I_2 = \frac{1}{J(\overline{A}, \phi A)} \frac{A(-1)}{q-1} \frac{g(\eta\overline{\chi})g(\overline{\eta}\overline{\chi})}{g(\phi\overline{\chi})g(\overline{\chi})}$$

$$\cdot \sum_{\varphi \in \widehat{\mathbb{F}_q^\times}} J(A\varphi, \overline{\varphi}) J(\overline{A}\varphi, \phi\overline{\varphi}) g(\phi\varphi\overline{\chi}) g(\varphi\overline{\chi}) g(\overline{\varphi}\chi\overline{\eta}) g(\overline{\varphi}\chi\eta)$$

$$\cdot \left(\frac{\varphi\chi(-1)}{q} + \frac{q-1}{q}\delta\left(\chi\overline{\varphi}\overline{\eta}\right)\right) \left(\frac{\varphi\chi(-1)}{q} + \frac{q-1}{q}\delta\left(\chi\overline{\varphi}\eta\right)\right)$$

$$= \frac{A(-1)}{q-1} \frac{1}{J(\overline{A}, \phi A)} \frac{g(\eta\overline{\chi})g(\overline{\eta}\overline{\chi})}{g(\phi\overline{\chi})g(\overline{\chi})} g(\eta)g(\phi\overline{\eta})$$

$$\cdot \sum_\varphi J(A\varphi, \overline{\varphi}) J(\overline{A}\varphi, \phi\overline{\varphi}) J(\phi\varphi\overline{\chi}, \overline{\eta}\chi\overline{\varphi}) J(\overline{\chi}\varphi, \chi\eta\overline{\varphi})$$

$$\cdot \left(\frac{1}{q^2} + \frac{q-1}{q^2}\delta\left(\chi\overline{\varphi}\overline{\eta}\right) + \frac{q-1}{q^2}\delta\left(\chi\overline{\varphi}\eta\right)\right).$$

The first delta term gives us

$$-\frac{A(-1)}{q^2} \frac{J(\overline{\eta}, \overline{\eta})}{J(\phi, \overline{\eta})} g(\eta\overline{\chi}) g(\overline{\eta}\overline{\chi}) g(A\chi\overline{\eta}) g(\overline{A}\chi\overline{\eta}) \frac{g(\eta\overline{\chi}) g(\phi\eta\overline{\chi})}{g(\phi\overline{\chi})g(\overline{\chi})}$$

$$= -\frac{A(-1)}{q^2} \frac{\overline{\eta}(4) J(\overline{\eta}, \overline{\eta})}{J(\phi, \overline{\eta})} J(\eta\overline{\chi}, \overline{\eta}\overline{\chi}) J(A\chi\overline{\eta}, \overline{A}\chi\overline{\eta}) g(\eta\chi^2) g(\overline{\eta}\overline{\chi}^2)$$

$$= -A(-1) \frac{J(A\chi\overline{\eta}, \overline{A}\chi\overline{\eta})}{J(\eta\chi, \overline{\eta}\chi)},$$

using (2.16). The second delta term contributes
$$-A(-1)\frac{J(A\eta\chi,\overline{A}\eta\chi)}{J(\eta\chi,\overline{\eta}\chi)}.$$
Therefore, we have
$$I = A(-1)\frac{J(A\chi,\overline{A}\chi)}{J(\eta\chi,\overline{\eta}\chi)}$$
$$+ \frac{\phi(-1)}{q^2}\frac{g(\eta)g(\phi\overline{\eta})}{J(\overline{A},\phi A)}\frac{g(\eta\overline{\chi})g(\overline{\eta}\chi)}{g(\overline{\chi})g(\phi\overline{\chi})}{}_4\mathbb{P}_3\begin{bmatrix} A & \overline{A} & \overline{\chi}\phi & \overline{\chi} \\ & \phi & \overline{\chi}\eta & \overline{\chi}\eta \end{bmatrix};1\Big].$$

We are now in the position to evaluate ${}_4\mathbb{P}_3\begin{bmatrix} A & \overline{A} & \overline{\chi}\phi & \overline{\chi} \\ & \phi & \overline{\chi}\eta & \overline{\chi}\eta \end{bmatrix};1\Big]$. For this purpose, we observe that for $x \neq 0$,
$${}_2\mathbb{P}_1\begin{bmatrix} A & \overline{A} \\ & \phi \end{bmatrix};\frac{1}{x}\Big]{}_2\mathbb{P}_1\begin{bmatrix} \eta & \overline{\eta} \\ & \phi \end{bmatrix};\frac{1}{x}\Big] = \frac{\phi(-1)}{q-1}\sum_{\chi}{}_4\mathbb{P}_3\begin{bmatrix} A & \overline{A} & \overline{\chi}\phi & \overline{\chi} \\ & \phi & \overline{\chi}\eta & \overline{\chi}\eta \end{bmatrix};1\Big]\overline{\chi}(x),$$

and by Proposition 8.1 and Corollary 8.14, the left hand side of the above equals
$$A\eta(x)J(\phi A,\phi A)J(\phi\eta,\phi\eta){}_2\mathbb{F}_1\begin{bmatrix} A & \phi A \\ & A^2 \end{bmatrix};x\Big]{}_2\mathbb{F}_1\begin{bmatrix} \eta & \phi\eta \\ & \overline{\eta} \end{bmatrix};x\Big]$$
$$= A\eta(x)J(\phi A,\phi A)J(\phi\eta,\phi\eta)$$
$$\cdot \left({}_2\mathbb{F}_1\begin{bmatrix} A\eta & \phi A\eta \\ & (A\eta)^2 \end{bmatrix};x\Big] + \eta(2x){}_2\mathbb{F}_1\begin{bmatrix} A\overline{\eta} & \phi A\overline{\eta} \\ & (A\overline{\eta})^2 \end{bmatrix};x\Big] - \delta(1-x)A\eta(4)\right).$$

From (2.16) and the relation
$$J(\phi,\phi\chi) = \phi(-1)J(\overline{\chi},\phi), \text{ if } \chi \neq \varepsilon,$$
we obtain
$$A\eta(x)J(\phi A,\phi A)J(\phi\eta,\phi\eta){}_2\mathbb{F}_1\begin{bmatrix} A\eta & \phi A\eta \\ & (A\eta)^2 \end{bmatrix};x\Big]$$
$$= A\eta(x)\phi(-1)\frac{J(\phi,\overline{A})J(\phi,\overline{\eta})}{J(\phi,\overline{A\eta})}{}_2\mathbb{P}_1\begin{bmatrix} A\eta & \phi A\eta \\ & A^2\overline{\eta} \end{bmatrix};x\Big]$$
$$= \frac{J(\phi,\overline{A})J(\phi,\overline{\eta})}{J(\phi,\overline{A\eta})}\frac{A(-1)}{q-1}\sum_{K\in\widehat{\mathbb{F}_q^\times}}J(A\eta\overline{K},K)J(\phi A\eta\overline{K},\overline{A}^2\eta K)\overline{K}A\eta(x)$$
$$= \frac{J(\phi,\overline{A})J(\phi,\overline{\eta})}{J(\phi,\overline{A\eta})}\frac{A(-1)}{q-1}\sum_{\chi\in\widehat{\mathbb{F}_q^\times}}J(\overline{\chi},A\eta\chi)J(\phi\overline{\chi},\overline{A}\eta\chi)\overline{\chi}(x),$$

$$A\eta(x)J(\phi A,\phi A)J(\phi\eta,\phi\eta)\eta(2x){}_2\mathbb{F}_1\begin{bmatrix} A\overline{\eta} & \phi A\overline{\eta} \\ & (A\overline{\eta})^2 \end{bmatrix};x\Big]$$
$$= \frac{J(\phi,\overline{A})J(\phi,\overline{\eta})}{J(\phi,\overline{A\eta})}\frac{A(-1)}{q-1}\sum_{\chi\in\widehat{\mathbb{F}_q^\times}}J(\overline{\chi},A\overline{\eta}\chi)J(\phi\overline{\chi},\overline{A}\eta\chi)\overline{\chi}(x),$$

and
$$A\eta(x)J(\phi A,\phi A)J(\phi\eta,\phi\eta)A\eta(4)\delta(1-x) = \frac{1}{q-1}J(\phi,\overline{A})J(\phi,\overline{\eta})\sum_{\chi\in\widehat{\mathbb{F}_q^\times}}\overline{\chi}(x).$$

Therefore, by comparing the coefficients for $\overline{\chi}$, we get
$$\phi(-1)\,_4\mathbb{P}_3\left[\begin{array}{cccc} A & \overline{A} & \overline{\chi}\phi & \overline{\chi} \\ & \phi & \overline{\chi}\eta_3 & \overline{\chi}\eta_3 \end{array};1\right]$$
$$= A(-1)J(\phi,\overline{A})J(\phi,\overline{\eta})\left(\frac{J(\overline{\chi},A\eta\chi)J(\phi\overline{\chi},\overline{A}\overline{\eta}\chi)}{J(\phi,\overline{A}\eta)}+\frac{J(\overline{\chi},A\overline{\eta}\chi)J(\phi\overline{\chi},\overline{A}\eta\chi)}{J(\phi,\overline{A}\eta)}\right)$$
$$-J(\phi,\overline{A})J(\phi,\overline{\eta}),$$

and when $\chi^2 \neq \varepsilon$, this is equal to
$$J(\phi,\overline{A})J(\phi,\overline{\eta})\left(\chi(4)J(A\eta\chi,\overline{A}\overline{\eta}\chi)+\chi(4)J(A\overline{\eta}\chi,\overline{A}\eta\chi)-1\right).$$

Our claim follows from these results and the fact that $\phi(-3) = 1$ when $p \equiv 1 \pmod{6}$. □

REMARK 9.27. Gessel and Stanton obtained in [**38**] many other interesting evaluation formulas. Many of them appear to have finite field analogues. Interested readers are encouraged to take a look.

CHAPTER 10

An application to Hypergeometric Abelian Varieties

In this chapter, we give an explicit application of the use of finite field formulas in computing the arithmetic invariants of hypergeometric varieties. Specifically, we use the finite field quadratic transformation from Theorem 9.4 to obtain the decomposition of a generically 4-dimensional abelian variety arising naturally from the generalized Legendre curve $y^{12} = x^9(1-x)^5(1-\lambda x)$.

In [**23**], based on [**103**] by Wolfart and [**8**] by Archinard, the authors use generalized Legendre curves (see §6.2) to construct families of 2-dimensional abelian varieties with quaternionic multiplication (QM) that are parametrized by Shimura curves associated with arithmetic triangle groups with compact fundamental domains. For example, the primitive part of the Jacobian varieties of (the smooth models of) $y^6 = x^4(1-x)^3(1-\lambda x)$ gives rise to a family of 2-dimensional abelian varieties parametrized by the Shimura curve associated with the arithmetic triangle group (3,6,6). Another construction of such a family of abelian varieties has been obtained by Petkova-Shiga [**78**] using Picard curves. In general the strategy in [**23**] yields computable families of generalized Legendre curves giving rise to hypergeometric abelian varieties of dimension larger than 2. For instance, the arithmetic triangle group (2,6,6) can be realized using the periods of the generalized Legendre curve $y^{12} = x^9(1-x)^5(1-\lambda x)$. (See Examples 3.5 and 6.2 for relations between the Legendre curves and the arithmetic triangle groups (3,6,6) and (2,6,6).) By the discussion in §6.2, for any $\lambda \in \mathbb{Q} \setminus \{0,1\}$, the primitive part of the Jacobian variety of the smooth model of $y^{12} = x^9(1-x)^5(1-\lambda x)$ is of dimension $\varphi(12) = 4$. It is natural to ask whether such a 4-dimensional abelian variety is simple or not. Meanwhile the two arithmetic groups (3,6,6) and (2,6,6) are commensurable, as seen in [**92**] by Takeuchi. Here, we will use Theorem 9.4, which is a quadratic formula for $_2\mathbb{F}_1$, to prove the following theorem.

THEOREM 10.1. *Let $\lambda \in \mathbb{Q}$ such that $\lambda \neq 0, \pm 1$. Let $J^{prim}_{\lambda,1}$ (resp. $J^{prim}_{\frac{-4\lambda}{(1-\lambda)^2},2}$) be the primitive part of the Jacobian variety of $y^6 = x^4(1-x)^3(1-\lambda x)$ (resp. $y^{12} = x^9(1-x)^5(1+\frac{4\lambda}{(1-\lambda)^2}x)$). Then $J^{prim}_{\frac{-4\lambda}{(1-\lambda)^2},2}$ is isogenous to $J^{prim}_{\lambda,1} \oplus J^{prim}_{\lambda,1}$ over some cyclotomic number field depending on λ.*

The proof is based on the following famous result of Faltings [**34**] (see §5, Kor. 1 to Satz 4).

THEOREM 10.2 (Faltings [**34**]). *Let A and B be abelian varieties over a number field L. Suppose that the corresponding ℓ-adic representations $\rho_{A,\ell} \simeq \rho_{B,\ell}$ as $\overline{\mathbb{Q}}_\ell[G_L]$-modules. Then A is isogenous to B.*

It is well-known that the $\rho_{A,\ell}$ and $\rho_{B,\ell}$ in Theorem 10.2 are semisimple and that Frobenius elements for $\rho_{A,\ell}$ and $\rho_{B,\ell}$ form a dense set of G_L. From this we have the following consequence.

COROLLARY 10.3. *If A and B are two abelian varieties over a number field L such that their corresponding Galois representations $\rho_{A,\ell}, \rho_{B,\ell}$ have the same trace for almost all Frobenius elements of G_L, then A is isogenous to B over L.*

PROOF OF THEOREM 10.1. For any fixed $\lambda \in \mathbb{Q}, \lambda \neq 0, \pm 1$, use $\{\rho_{1,\ell}\}$ (resp. $\{\rho_{2,\ell}\}$) to denote the compatible family of 4-dimensional (resp. 8-dimensional) Galois representations of $G_\mathbb{Q}$ arising from $J_{\lambda,1}^{\text{prim}}$ (resp. $J_{\frac{-4\lambda}{(1-\lambda)^2},2}^{\text{prim}}$) respectively. By Corollary 10.3 we only need to find a finite extension L of $\mathbb{Q}$ such that the traces of $\rho_{2,\ell}|_{G_L}$ and $(\rho_{1,\ell} \oplus \rho_{1,\ell})|_{G_L}$ agree at almost all Frobenius elements of G_L.

Let $\mathfrak{p}$ be a good prime ideal of the ring of integers of $\mathbb{Q}(\zeta_{12})$ with residue field of size q. Note that $q \equiv 1 \pmod{12}$. Then the trace of the Frobenius element $\text{Frob}_\mathfrak{p}$ under $\{\rho_{1,\ell}\}$ and $\{\rho_{2,\ell}\}$ can be computed using (6.8). All the fractions on the right sides below are negative of those given by (6.8). As the sets we are summing over are stable under negation, our formulas are correct.

$$(10.1) \qquad \text{Tr}\rho_{1,\ell}(\text{Frob}_\mathfrak{p}) = -\sum_{m=1,5} {}_2\mathbb{P}_1 \begin{bmatrix} \iota_\mathfrak{p}(\frac{m}{6}) & \iota_\mathfrak{p}(\frac{2m}{6}) \\ & \iota_\mathfrak{p}(\frac{-m}{6}) \end{bmatrix} ; \lambda; q \end{bmatrix},$$

and

$$(10.2) \qquad \text{Tr}\rho_{2,\ell}(\text{Frob}_\mathfrak{p}) = -\sum_{m=1,5,7,11} {}_2\mathbb{P}_1 \begin{bmatrix} \iota_\mathfrak{p}(\frac{m}{12}) & \iota_\mathfrak{p}(\frac{3m}{12}) \\ & \iota_\mathfrak{p}(\frac{-2m}{12}) \end{bmatrix} ; \frac{-4\lambda}{(1-\lambda)^2}; q \end{bmatrix},$$

respectively. (For the notation $\iota_\mathfrak{p}(\frac{a}{b})$, see Definition 5.9.) Assume $\lambda \neq 0, \pm 1$ in the residue field. We now let $\eta_{12} = \iota_\mathfrak{p}(\frac{m}{12})$ with $(m, 12) = 1$ be any primitive multiplicative character of order 12. Then $\eta_{12}^6 = \phi$ is the unique quadratic character. By applying Theorem 9.4 with $D = \eta_{12}$, $C = \eta_{12}^2$ and $B = \eta_{12}^4$ in the ${}_2\mathbb{F}_1$ expression, invoking the commutativity in §4.3, and using the defining formula (2.12) for ${}_2\mathbb{P}_1$, we see for $\lambda \neq 0, \pm 1$ that

$${}_2\mathbb{P}_1 \begin{bmatrix} \eta_{12}^2 & \eta_{12}^4 \\ & \overline{\eta}_{12}^2 \end{bmatrix} ; \lambda \end{bmatrix} = \overline{\eta}_{12}^2(1-\lambda) \frac{J(\eta_{12}^4, \phi)}{J(\eta_{12}^3, \overline{\eta}_{12}^5)} {}_2\mathbb{P}_1 \begin{bmatrix} \eta_{12} & \eta_{12}^3 \\ & \overline{\eta}_{12}^2 \end{bmatrix} ; \frac{-4\lambda}{(1-\lambda)^2} \end{bmatrix}.$$

Replacing all characters by their 5th powers gives

$${}_2\mathbb{P}_1 \begin{bmatrix} \overline{\eta}_{12}^2 & \overline{\eta}_{12}^4 \\ & \eta_{12}^2 \end{bmatrix} ; \lambda \end{bmatrix} = \eta_{12}^2(1-\lambda) \frac{J(\overline{\eta}_{12}^4, \phi)}{J(\overline{\eta}_{12}^3, \overline{\eta}_{12})} {}_2\mathbb{P}_1 \begin{bmatrix} \eta_{12}^5 & \eta_{12}^3 \\ & \eta_{12}^2 \end{bmatrix} ; \frac{-4\lambda}{(1-\lambda)^2} \end{bmatrix}.$$

Let $R(\eta_{12}) := \frac{J(\eta_{12}^4, \phi)}{J(\eta_{12}^3, \overline{\eta}_{12}^5)}$. Next, we will argue that $R(\eta_{12})$ is a root of unity. As $R(\eta_{12})$ and $R(\eta_{12}^5)$ are Galois conjugates in $\mathbb{Q}(\zeta_{12})$ it suffices to show $R(\eta_{12}^5)$ is a root of unity.

First we note that by the reflection formula (2.10), we can show that

$$g(\eta_{12}^k)g(\eta_{12}^{12-k}) = \eta_{12}(-1)^k q.$$

Thus further using (2.10) we see that

10. AN APPLICATION TO HYPERGEOMETRIC ABELIAN VARIETIES

$$(10.3) \quad R(\eta_{12}^5) = \frac{g(\eta_{12}^8)g(\phi)}{g(\eta_{12}^3)g(\eta_{12}^{11})} = \frac{g(\phi)g(\eta_{12}^8)g(\eta_{12})}{\eta_{12}(-1)qg(\eta_{12}^3)}$$

$$= \eta_{12}(-1)\frac{g(\phi)g(\eta_{12})}{g(\eta_{12}^3)g(\eta_{12}^4)} = \eta_{12}(-1)\frac{g(\phi)g(\eta_{12})}{g(\eta_{12}^3)g(\eta_{12}^4)}\frac{g(\eta_{12}^7)g(\eta_{12}^5)}{\eta_{12}(-1)q}.$$

Now letting $A = \eta_{12}$ in the multiplication formula (2.12) then using (2.10) and (2.13), we obtain

$$\eta_{12}(4)R(\eta_{12}^5) = \frac{g(\phi)^2 g(\eta_{12}^2)g(\eta_{12}^5)}{qg(\eta_{12}^3)g(\eta_{12}^4)} = \frac{g(\eta_{12}^2)g(\eta_{12}^5)}{g(\eta_{12}^3)g(\eta_{12}^4)} = \frac{J(\eta_{12}^2, \eta_{12}^5)}{J(\eta_{12}^3, \eta_{12}^4)}.$$

By Example 5.15, the value of $J(\eta_{12}^2, \eta_{12}^5)/J(\eta_{12}^3, \eta_{12}^4)$ is also a root of unity. From a more global view point following Weil's Theorem 5.13, we know now the corresponding Grössencharacter $\mathcal{J}_{(\frac{1}{3}, \frac{1}{2})}/\mathcal{J}_{(\frac{1}{4}, \frac{7}{12})}$ has finite image. By class field theory, it corresponds to a finite order character ψ. Hence there is a natural number M, corresponding to the intersections of the kernels of $\left(\frac{-(1-\lambda)^2}{\cdot}\right)_{12}$ and ψ, such that for each good prime $\mathfrak{p}$ whose residue field has cardinality 1 modulo M, then $\eta_{12}^2(1-\lambda) = 1, \eta_{12}(-1) = 1$, and $\frac{J(\eta_{12}^4, \phi)}{J(\eta_{12}^3, \overline{\eta}_{12}^5)} = 1$.

Set $L = \mathbb{Q}(\zeta_{12M})$. Then for each good prime ideal $\mathfrak{p}$ of $\mathcal{O}_L$, its residue field has size $q \equiv 1 \pmod{12M}$. In particular, letting $\eta_{12} = \iota_{\mathfrak{p}}(\frac{1}{12})$ and $\eta_{12} = \iota_{\mathfrak{p}}(\frac{7}{12})$, respectively, in the above equations (10.1) and (10.2), respectively, yields

$$\text{Tr}\rho_{2,\ell}|_{G_L}(\text{Frob}_{\mathfrak{p}}) = 2\text{Tr}\rho_{1,\ell}|_{G_L}(\text{Frob}_{\mathfrak{p}}).$$

The claim then follows from Corollary 10.3. □

CHAPTER 11

Open Questions and Concluding Remarks

We have seen now various ways in which the finite field hypergeometric series defined in (4.9) can be leveraged. By slightly changing the definitions given in [41] by Greene and [70] by McCarthy, we are able to accomplish our goals of aligning with the underlining geometry and matching the classical setting as closely as possible. This gives us a systematic method for converting many classical results to the finite field setting, with the benefit of predictions motivated by the Galois perspective.

In particular, we have used our methods to prove finite field analogues of numerous classical formulas including 9 quadratic or higher transformation formulas, 11 evaluation formulas, and 3 algebraic identities, among other formulas.

In addition, we will discuss some numeric observations in this chapter.

11.1. Numeric observations

Numerically, we have observed finite field analogues of the algebraic transformations of $_2F_1$-hypergeometric series mentioned in work of the fifth author and Yang [94] using Magma. We have the following conjecture, for which we use the convention that a rational function $f(x) = p(x)/q(x)$ is said to take value ∞ at $x = a$ if $q(a) = 0$ but $p(a) \neq 0$.

CONJECTURE 11.1. *Let $p \equiv 1 \pmod{24}$, α a root of $x^2 + 3$ and β a root of $x^2 + 2$ in $\mathbb{F}_p$, and η_{24} a primitive multiplicative character of order 24. Define*

$$f(z) := \frac{12\alpha z(1-z)^2(1-9z^2)}{(1+\alpha z)^6},$$

$$g(z) := -\frac{4(1+\beta)^4 z(1+(4\beta-7)z^2/3)^4}{(1+z)(1-3z)(1+(4+2\beta)z-(1+2\beta)z^2)^4},$$

and assume that $z \in \mathbb{F}_p$ satisfies $f(z), g(z) \neq 0, 1, \infty$. Then,

$$\eta_{24}^3((1+z)(1-3z))\overline{\eta}_{24}^6(1+\alpha z) \, _2F_1\left[\begin{matrix}\eta_{24}^5 & \eta_{24}^9 \\ & \overline{\eta}_{24}^6\end{matrix}; f(z)\right]$$

$$\stackrel{?}{=} \phi(1+(4+2\beta)z-(1+2\beta)z^2) \, _2F_1\left[\begin{matrix}\eta_{24}^3 & \eta_{24}^9 \\ & \overline{\eta}_{24}^6\end{matrix}; g(z)\right].$$

Conjecture 11.1 is a finite field analogue of the following transformation[1] due to the fifth author and Yang [94].

---

[1] We note that the left hand side of the second formula in [94, Theorem 1] contains a typo. We are stating the corrected version here.

11. OPEN QUESTIONS AND CONCLUDING REMARKS

THEOREM 11.2. **[94**, Theorem 1**]** *Let α be a root of x^2+3, β a root of x^2+2, and define*

$$f(z) := \frac{12\alpha z(1-z)^2(1-9z^2)}{(1+\alpha z)^6}$$

$$g(z) := -\frac{4(1+\beta)^4 z(1+(4\beta-7)z^2/3)^4}{(1+z)(1-3z)(1+(4+2\beta)z-(1+2\beta)z^2)^4}.$$

Then,

$$\frac{(1+z)^{1/8}(1-3z)^{1/8}}{(1+\alpha z)^{5/4}} {}_2F_1\left[\begin{array}{cc}\frac{5}{24} & \frac{3}{8} \\ & \frac{3}{4}\end{array}; f(z)\right]$$
$$= (1+(4+2\beta)z-(1+2\beta)z^2)^{-1/2} {}_2F_1\left[\begin{array}{cc}\frac{1}{8} & \frac{3}{8} \\ & \frac{3}{4}\end{array}; g(z)\right],$$

and

$$\frac{(1-z)^{1/4}(1+3z)^{1/4}(1+z)^{5/8}(1-3z)^{5/8}}{(1+\alpha z)^{11/4}} {}_2F_1\left[\begin{array}{cc}\frac{11}{24} & \frac{5}{8} \\ & \frac{5}{4}\end{array}; f(z)\right]$$
$$= \frac{1+(4\beta-7)z^2/3}{(1+(4+2\beta)z-(1+2\beta)z^2)^{3/2}} {}_2F_1\left[\begin{array}{cc}\frac{3}{8} & \frac{5}{8} \\ & \frac{5}{4}\end{array}; g(z)\right].$$

By §3.2, the projective monodromy groups of the differential equations satisfied by ${}_2F_1\left[\begin{array}{cc}\frac{5}{24} & \frac{3}{8} \\ & \frac{3}{4}\end{array}; z\right]$ and ${}_2F_1\left[\begin{array}{cc}\frac{11}{24} & \frac{5}{8} \\ & \frac{5}{4}\end{array}; z\right]$ are both triangle groups. In fact they are isomorphic to two commensurable arithmetic triangle groups $(4,6,6)$ and $(4,4,4)$ respectively. In [**94**], the authors obtain the transformations by interpreting hypergeometric series as modular forms on Shimura curves. In the case of the above theorem, all the modular forms for the arithmetic triangle group $(4,6,6)$ can be expressed in terms of

$${}_2F_1\left[\begin{array}{cc}\frac{5}{24} & \frac{3}{8} \\ & \frac{3}{4}\end{array}; t\right] \quad \text{and} \quad t^{1/4}\, {}_2F_1\left[\begin{array}{cc}\frac{11}{24} & \frac{5}{8} \\ & \frac{5}{4}\end{array}; t\right],$$

and for the Shimura curve associated to $(4,4,4)$ can be expressed in terms of

$${}_2F_1\left[\begin{array}{cc}\frac{1}{8} & \frac{3}{8} \\ & \frac{3}{4}\end{array}; u\right] \quad \text{and} \quad u^{1/4}\, {}_2F_1\left[\begin{array}{cc}\frac{3}{8} & \frac{5}{8} \\ & \frac{5}{4}\end{array}; u\right],$$

where both t and u are suitable meromorphic modular functions. All of them can be regarded as modular forms for the intersection group Γ of $(4,6,6)$ and $(4,4,4)$, which is an arithmetic group generated by six elliptic elements of order 4 with a single relation. The algebraic transformations between these hypergeometric series come from the identities among the modular forms with respect to Γ.

We have also observed numerically a finite field version of the second formula in Theorem 11.2. Namely,

$$(11.1) \quad \eta_{24}^6((1-z)(1+3z)(1+\alpha z))\overline{\eta}_{24}^9((1+z)(1-3z))\, {}_2\mathbb{F}_1\left[\begin{array}{cc}\eta_{24}^{11} & \overline{\eta}_{24}^9 \\ & \eta_{24}^6\end{array}; f(z)\right]$$
$$\stackrel{?}{=} \phi(1+(4+2\beta)z-(1+2\beta)z^2)\, {}_2\mathbb{F}_1\left[\begin{array}{cc}\eta_{24}^9 & \overline{\eta}_{24}^9 \\ & \eta_{24}^6\end{array}; g(z)\right].$$

11.1. NUMERIC OBSERVATIONS

Equation (11.1) is equivalent to Conjecture 11.1, which can be seen by applying the first item of Proposition 8.5 and the multiplication formula in Theorem 2.7. In fact this is predicted by the Galois perspective as the finite field versions correspond to the traces of two dimensional Galois representations.

Furthermore, we have the following conjectures, based on numerical evidence computed using Magma, for $\mathbb{F}_q$ analogues of the algebraic transformations stated in Theorems 2, 4, 5 of [**94**].

CONJECTURE 11.3. *Let η_{20} be a primitive character of order* 20 *and define*
$$f(z) := \frac{64z(1-z-z^2)^5}{(1-z^2)(1+4z-z^2)^5}.$$
Then when $f(z) \neq 0, 1, \infty$,

$$_2\mathbb{F}_1 \begin{bmatrix} \eta_{20} & \eta_{20}^5 \\ & \overline{\eta}_{20}^4 \end{bmatrix} ; f(z) \end{bmatrix} = \eta_{20}(1-z^2)\eta_{20}^5(1+4z-z^2) \, _2\mathbb{F}_1 \begin{bmatrix} \eta_{20}^6 & \eta_{20}^8 \\ & \overline{\eta}_{20}^2 \end{bmatrix} ; z^2 \end{bmatrix}.$$

Conjecture 11.3 is an analogue of the following theorem.

THEOREM 11.4 (Theorem 2, [**94**]). *For $z \in \mathbb{C}$ such that both sides converge,*

$$_2F_1 \begin{bmatrix} \frac{1}{20} & \frac{1}{4} \\ & \frac{4}{5} \end{bmatrix} ; \frac{64z(1-z-z^2)^5}{(1-z^2)(1+4z-z^2)^5} \end{bmatrix}$$
$$= (1-z^2)^{1/20}(1+4z-z^2)^{1/4} \, _2F_1 \begin{bmatrix} \frac{3}{10} & \frac{2}{5} \\ & \frac{9}{10} \end{bmatrix} ; z^2 \end{bmatrix};$$

$$(1-z-z^2) \, _2F_1 \begin{bmatrix} \frac{9}{20} & \frac{1}{4} \\ & \frac{6}{5} \end{bmatrix} ; \frac{64z(1-z-z^2)^5}{(1-z^2)(1+4z-z^2)^5} \end{bmatrix}$$
$$= (1-z^2)^{1/4}(1+4z-z^2)^{5/4} \, _2F_1 \begin{bmatrix} \frac{1}{2} & \frac{2}{5} \\ & \frac{11}{10} \end{bmatrix} ; z^2 \end{bmatrix}.$$

Next, we have the following conjecture.

CONJECTURE 11.5. *Let η_6 be a primitive character of order* 6 *and A a character with $A^6 \neq \varepsilon$. Then when $z \neq \pm 1, \pm 3$,*

$$A\eta_6(1+z)A^3\phi(1-z/3) \, _2\mathbb{F}_1 \begin{bmatrix} A^2\eta_6^2 & A\eta_6^2 \\ & A^3 \end{bmatrix} ; z^2 \end{bmatrix}$$
$$= \, _2\mathbb{F}_1 \begin{bmatrix} A\eta_6 & A\phi \\ & A^2 \end{bmatrix} ; \frac{16z^3}{(1+z)(3-z)^3} \end{bmatrix}.$$

Conjecture 11.5 is an analogue of the following theorem.

THEOREM 11.6 (Theorem 4, [**94**]). *For a rational number a such that neither $3a + 1$ nor $2a + 1$ is a nonpositive integer, in a neighborhood of $z = 0$,*

$$(1+z)^{a+1/6}(1-z/3)^{3a+1/2} \, _2F_1 \begin{bmatrix} 2a+\frac{1}{3} & a+\frac{1}{3} \\ & 3a+1 \end{bmatrix} ; z^2 \end{bmatrix}$$
$$= \, _2F_1 \begin{bmatrix} a+\frac{1}{6} & a+\frac{1}{2} \\ & 2a+1 \end{bmatrix} ; \frac{16z^3}{(1+z)(3-z)^3} \end{bmatrix}.$$

Our final conjecture is stated below.

CONJECTURE 11.7. *Let η_{12} be a primitive character of order 12 and A a character with $A^{12} \neq \varepsilon$. Then when neither $-\frac{27z^2(1-z)}{1-9z}$ nor $-\frac{64z^3}{(1-z)^3(1-9z)}$ equals 0, 1, or ∞,*

$$A^9 \eta_{12}^9 (1-z) \,_2F_1 \left[\begin{matrix} A^4 \eta_{12}^4 & A^2 \eta_{12}^4 \\ & A^6 \end{matrix} ; -\frac{27z^2(1-z)}{1-9z} \right]$$
$$= A\eta_{12}(1-9z) \,_2F_1 \left[\begin{matrix} A^3 \eta_{12}^3 & A\eta_{12}^3 \\ & A^4 \end{matrix} ; -\frac{64z^3}{(1-z)^3(1-9z)} \right].$$

Conjecture 11.7 is an analogue of the following theorem.

THEOREM 11.8 (Theorem 5, [**94**]). *For a real number a such that neither $6a+1$ nor $4a+1$ is a non-positive integer, then in a neighborhood of $z = 0$,*

$$(1-z)^{9a+3/4} \,_2F_1 \left[\begin{matrix} 4a+\frac{1}{3} & 2a+\frac{1}{3} \\ & 6a+1 \end{matrix} ; -\frac{27z^2(1-z)}{1-9z} \right]$$
$$= (1-9z)^{a+1/12} \,_2F_1 \left[\begin{matrix} 3a+\frac{1}{4} & a+\frac{1}{4} \\ & 4a+1 \end{matrix} ; -\frac{64z^3}{(1-z)^3(1-9z)} \right].$$

Note that it is unclear whether the transformation theorems of [**94**] cited here satisfy the (∗) condition or not as they are proved using automorphic forms and the proofs do not have direct finite field translations.

CHAPTER 12

Appendix

In this appendix we address a few remaining topics not otherwise discussed in the bulk of this work. In §12.1 and §12.2 we offer alternate proofs and/or outlines of proofs for some classical results which demonstrate that the ($*$) condition is satisfied.

12.1. Bailey $_3F_2$ cubic transforms

Here we show that (9.18) and (9.19) satisfy the ($*$) condition. See Bailey [10] for the original proofs of these transformations.

First observe that from (2.2) we have

$$(12.1) \quad \binom{-a-3i}{n-i} = (-1)^n \frac{(a)_n(-n)_i(a+n)_{2i}}{n!(a)_{3i}}$$
$$= (-1)^n \frac{(a)_n(-n)_i((a+n)/2)_i((a+n+1)/2)_i}{n!(a/3)_i((a+1)/3)_i((a+2)/3)_i} \left(\frac{4}{27}\right)^i,$$

$$(12.2) \quad \binom{-a-3i}{n-2i} = (-1)^n \frac{(a)_n(a+n)_i(-n)_{2i}}{n!(a)_{3i}}$$
$$= (-1)^n \frac{(a)_n(a+n)_i(-n/2)_i((1-n)/2)_i}{n!(a/3)_i((a+1)/3)_i((a+2)/3)_i} \left(\frac{4}{27}\right)^i.$$

PROOF OF (9.19). Note that the left hand side of (9.19) is

$$\sum_{i\geq 0} \frac{\left(\frac{a}{3}\right)_i \left(\frac{a+1}{3}\right)_i \left(\frac{a+2}{3}\right)_i}{i!(b)_i \left(a+\frac{3}{2}-b\right)_i} \left(\frac{-27}{4}\right)^i x^i (1-x)^{-a-3i}$$
$$= \sum_{i\geq 0} \frac{\left(\frac{a}{3}\right)_i \left(\frac{a+1}{3}\right)_i \left(\frac{a+2}{3}\right)_i}{i!(b)_i \left(a+\frac{3}{2}-b\right)_i} \left(\frac{-27}{4}\right)^i x^i \sum_{k\geq 0} \binom{-a-3i}{k} (-x)^k$$
$$\stackrel{n=i+k}{=} \sum_{n,i\geq 0, n\geq i} \frac{\left(\frac{a}{3}\right)_i \left(\frac{a+1}{3}\right)_i \left(\frac{a+2}{3}\right)_i}{i!(b)_i \left(a+\frac{3}{2}-b\right)_i} \left(\frac{27}{4}\right)^i \binom{-a-3i}{n-i} (-x)^n$$
$$\stackrel{(12.1)}{=} \sum_{n\geq 0} \frac{(a)_n}{n!} {}_3F_2\left[\begin{matrix}-n & (a+n)/2 & (a+n+1)/2 \\ & b & a+3/2-b\end{matrix};1\right] x^n.$$

113

Applying the Pfaff-Saalschütz formula (3.15), this becomes

$$\sum_{n\geq 0} \frac{(a)_n}{n!} \frac{(b-(a+n)/2)_n(b-(a+n+1)/2)_n}{(b)_n(b-a-n-1/2)_n} x^n$$

$$\stackrel{(2.8)}{=} \sum_n \frac{(a)_n}{n!} \frac{(2b-a-n-1)_{2n}}{(b)_n(b-a-n-1/2)_n} \left(\frac{x}{4}\right)^n$$

$$= \sum_n \frac{(a)_n \Gamma(2b-a+n-1)\Gamma(b-a-n-1/2)}{n!(b)_n \Gamma(2b-a-n-1)\Gamma(b-a-1/2)} \left(\frac{x}{4}\right)^n$$

$$= \sum_n \frac{(a)_n(2b-a-1)_n \Gamma(2b-a-1)\Gamma(b-a-n-1/2)}{n!(b)_n \Gamma(2b-a-n-1)\Gamma(b-a-1/2)} \left(\frac{x}{4}\right)^n$$

$$\stackrel{\text{Thm.2.4}}{=} \sum_n \frac{(a)_n(2b-a-1)_n \Gamma(2-2b+a+n)\Gamma(3/2-b+a)}{n!(b)_n \Gamma(2-2b+a)\Gamma(3/2-b+a+n)} \left(\frac{x}{4}\right)^n,$$

which agrees with the right hand side of (9.19). □

To prove (9.18) we note that one of $-n/2$ and $(1-n)/2$ is a nonpositive integer when $n \geq 0$. Thus using (12.2) the proof of (9.18) follows similarly to that of (9.19).

12.2. A proof of a formula by Gessel and Stanton

Recall that equation (9.26) states that

$${}_2F_1\left[\begin{matrix} a & -a \\ & \frac{1}{2} \end{matrix}; \frac{27x(1-x)^2}{4}\right] = {}_2F_1\left[\begin{matrix} 3a & -3a \\ & \frac{1}{2} \end{matrix}; \frac{3x}{4}\right].$$

We give a proof here which demonstrates that (9.26) satisfies the (∗) condition.

PROOF OF (9.26). Note that using the inversion formula (9.5) one has

$$(12.3) \qquad \binom{2n-2i}{i} = \frac{(-1)^i}{i!}\frac{(-2n)_{3i}}{(-2n)_{2i}} = (-1)^i \frac{27^i}{4^i \cdot i!} \frac{(\frac{-2n}{3})_i(\frac{1-2n}{3})_i(\frac{2-2n}{3})_i}{(-n)_i(-n+\frac{1}{2})_i}.$$

Observe that

$${}_2F_1\left[\begin{matrix} a & -a \\ & \frac{1}{2} \end{matrix}; \frac{27x(1-x)^2}{4}\right] = \sum_{k\geq 0} \frac{(a)_k(-a)_k}{(1)_k(\frac{1}{2})_k}\left(\frac{27x}{4}\right)^k \sum_{i\geq 0}\binom{2k}{i}(-x)^i$$

$$\stackrel{n=k+i}{=} \sum_{n,i\geq 0} \frac{(a)_{n-i}(-a)_{n-i}}{(1)_{n-i}(\frac{1}{2})_{n-i}}\left(\frac{27}{4}\right)^{n-i}\binom{2n-2i}{i}(-1)^i x^n$$

$$= \sum_{n\geq 0} \left(\frac{27}{4}\right)^n \frac{(a)_n(-a)_n}{(1)_n(\frac{1}{2})_n}\left(\sum_i \frac{(-n)_i(\frac{1}{2}-n)_i}{(1-a-n)_i(1+a-n)_i}\left(-\frac{27}{4}\right)^{-i}\binom{2n-2i}{i}\right)x^n.$$

By (12.3), the above equals

$$\sum_{n\geq 0} \frac{(a)_n(-a)_n}{(1)_n(\frac{1}{2})_n} {}_3F_2\left[\begin{matrix} \frac{-2n}{3} & \frac{1-2n}{3} & \frac{2-2n}{3} \\ 1+a-n & 1-a-n \end{matrix}; 1\right]\left(\frac{27x}{4}\right)^n$$

$$\stackrel{(3.16)}{=} \sum_{n\geq 0} \frac{(a)_n(-a)_n}{(1)_n(\frac{1}{2})_n} \frac{\Gamma(a+\frac{n+2}{3})\Gamma(a+\frac{n+1}{3})\Gamma(a+\frac{n}{3})\Gamma(1+a-n)}{\Gamma(1+a-\frac{n}{3})\Gamma(a+\frac{2-n}{3})\Gamma(a+\frac{1-n}{3})\Gamma(a+n)}\left(\frac{27x}{4}\right)^n.$$

12.2. A PROOF OF A FORMULA BY GESSEL AND STANTON

By Theorem 2.6, we thus have

$$\begin{aligned}
{}_2F_1\left[\begin{matrix}a & -a \\ & \frac{1}{2}\end{matrix}\,;\,\frac{27x(1-x)^2}{4}\right] &= 3\sum_{n\geq 0}\frac{(a)_n(-a)_n}{(1)_n(\frac{1}{2})_n}\frac{\Gamma(3a+n)\Gamma(1+a-n)}{\Gamma(1+3a-n)\Gamma(a+n)}\left(\frac{3x}{4}\right)^n \\
&= 3\sum_{n\geq 0}\frac{(3a)_n(-a)_n}{(1)_n(\frac{1}{2})_n}\frac{\Gamma(1+a-n)\Gamma(3a)}{\Gamma(1+3a-n)\Gamma(a)}\left(\frac{3x}{4}\right)^n \\
&= \sum_{n\geq 0}\frac{(3a)_n(-a)_n}{(1)_n(\frac{1}{2})_n}\frac{\Gamma(1+a-n)\Gamma(3a+1)}{\Gamma(1+3a-n)\Gamma(a+1)}\left(\frac{3x}{4}\right)^n \\
&\stackrel{\text{Thm.2.4}}{=} \sum_{n\geq 0}\frac{(3a)_n(-a)_n}{(1)_n(\frac{1}{2})_n}\frac{\Gamma(-3a+n)\Gamma(-a)}{\Gamma(-3a)\Gamma(-a+n)}\left(\frac{3x}{4}\right)^n \\
&= {}_2F_1\left[\begin{matrix}3a & -3a \\ & \frac{1}{2}\end{matrix}\,;\,\frac{3x}{4}\right]
\end{aligned}$$

as desired. □

Bibliography

[1] A. Adolphson and S. Sperber, *On twisted exponential sums*, Math. Ann. **290** (1991), no. 4, 713–726, DOI 10.1007/BF01459269. MR1119948

[2] A. Adolphson and S. Sperber, *Twisted exponential sums and Newton polyhedra*, J. Reine Angew. Math. **443** (1993), 151–177. MR1241131

[3] S. Ahlgren and K. Ono, *A Gaussian hypergeometric series evaluation and Apéry number congruences*, J. Reine Angew. Math. **518** (2000), 187–212, DOI 10.1515/crll.2000.004. MR1739404

[4] S. Ahlgren, *Gaussian hypergeometric series and combinatorial congruences*, Symbolic computation, number theory, special functions, physics and combinatorics (Gainesville, FL, 1999), Dev. Math., vol. 4, Kluwer Acad. Publ., Dordrecht, 2001, pp. 1–12, DOI 10.1007/978-1-4613-0257-5_1. MR1880076

[5] S. Ahlgren, K. Ono, and D. Penniston, *Zeta functions of an infinite family of $K3$ surfaces*, Amer. J. Math. **124** (2002), no. 2, 353–368. MR1890996

[6] G. E. Andrews, R. Askey, and R. Roy, *Special functions*, Encyclopedia of Mathematics and its Applications, vol. 71, Cambridge University Press, Cambridge, 1999, DOI 10.1017/CBO9781107325937. MR1688958

[7] G. E. Andrews and D. W. Stanton, *Determinants in plane partition enumeration*, European J. Combin. **19** (1998), no. 3, 273–282, DOI 10.1006/eujc.1997.0184. MR1621001

[8] N. Archinard, *Hypergeometric abelian varieties*, Canad. J. Math. **55** (2003), no. 5, 897–932, DOI 10.4153/CJM-2003-037-4. MR2005278

[9] N. Archinard, *Exceptional sets of hypergeometric series*, J. Number Theory **101** (2003), no. 2, 244–269, DOI 10.1016/S0022-314X(03)00042-8. MR1989887

[10] W. N. Bailey, *Products of Generalized Hypergeometric Series*, Proc. London Math. Soc. (2) **28** (1928), no. 4, 242–254, DOI 10.1112/plms/s2-28.1.242. MR1575853

[11] W. N. Bailey, *Generalized hypergeometric series*, Cambridge Tracts in Mathematics and Mathematical Physics, No. 32, Stechert-Hafner, Inc., New York, 1964. MR0185155

[12] F. Baldassarri and B. Dwork, *On second order linear differential equations with algebraic solutions*, Amer. J. Math. **101** (1979), no. 1, 42–76, DOI 10.2307/2373938. MR527825

[13] R. Barman and G. Kalita, *Hypergeometric functions and a family of algebraic curves*, Ramanujan J. **28** (2012), no. 2, 175–185, DOI 10.1007/s11139-011-9345-7. MR2925173

[14] B. C. Berndt, R. J. Evans, and K. S. Williams, *Gauss and Jacobi sums*, Canadian Mathematical Society Series of Monographs and Advanced Texts, John Wiley & Sons, Inc., New York, 1998. A Wiley-Interscience Publication. MR1625181

[15] F. Beukers. Notes of differential equations and hypergeometric functions. unpublished notes.

[16] F. Beukers, H. Cohen, and A. Mellit, *Finite hypergeometric functions*, Pure Appl. Math. Q. **11** (2015), no. 4, 559–589, DOI 10.4310/PAMQ.2015.v11.n4.a2. MR3613122

[17] F. Beukers and G. Heckman, *Monodromy for the hypergeometric function $_nF_{n-1}$*, Invent. Math. **95** (1989), no. 2, 325–354, DOI 10.1007/BF01393900. MR974906

[18] J. M. Borwein and P. B. Borwein, *A cubic counterpart of Jacobi's identity and the AGM*, Trans. Amer. Math. Soc. **323** (1991), no. 2, 691–701, DOI 10.2307/2001551. MR1010408

[19] J. M. Borwein and P. B. Borwein, *Pi and the AGM*, Canadian Mathematical Society Series of Monographs and Advanced Texts, vol. 4, John Wiley & Sons, Inc., New York, 1998. A study in analytic number theory and computational complexity; Reprint of the 1987 original; A Wiley-Interscience Publication. MR1641658

[20] S. Chisholm, A. Deines, L. Long, G. Nebe, and H. Swisher. p−adic analogues of ramanujan type formulas for $1/\pi$. *Mathematics*, 1(1):9–30, 2013.

[21] D. V. Chudnovsky and G. V. Chudnovsky, *Approximations and complex multiplication according to Ramanujan*, Ramanujan revisited (Urbana-Champaign, Ill., 1987), Academic Press, Boston, MA, 1988, pp. 375–472. MR938975

[22] H. Cohen, *Number theory. Vol. I. Tools and Diophantine equations*, Graduate Texts in Mathematics, vol. 239, Springer, New York, 2007. MR2312337

[23] A. Deines, J. G. Fuselier, L. Long, H. Swisher, and F.-T. Tu, *Generalized Legendre curves and quaternionic multiplication*, J. Number Theory **161** (2016), 175–203, DOI 10.1016/j.jnt.2015.04.019. MR3435724

[24] A. Deines, J. G. Fuselier, L. Long, H. Swisher, and F.-T. Tu, *Hypergeometric series, truncated hypergeometric series, and Gaussian hypergeometric functions*, Directions in number theory, Assoc. Women Math. Ser., vol. 3, Springer, [Cham], 2016, pp. 125–159, DOI 10.1007/978-3-319-30976-7_5. MR3596579

[25] A, Sutherland, *Modular Polynomials*, https://math.mit.edu/~drew/ClassicalModPolys.html

[26] R. Evans and J. Greene, *Clausen's theorem and hypergeometric functions over finite fields*, Finite Fields Appl. **15** (2009), no. 1, 97–109, DOI 10.1016/j.ffa.2008.09.001. MR2468995

[27] R. Evans and J. Greene, *Evaluations of hypergeometric functions over finite fields*, Hiroshima Math. J. **39** (2009), no. 2, 217–235. MR2543651

[28] R. Evans and J. Greene, *A quadratic hypergeometric $_2F_1$ transformation over finite fields*, Proc. Amer. Math. Soc. **145** (2017), no. 3, 1071–1076, DOI 10.1090/proc/13303. MR3589307

[29] R. J. Evans, *Identities for products of Gauss sums over finite fields*, Enseign. Math. (2) **27** (1981), no. 3-4, 197–209 (1982). MR659148

[30] R. J. Evans, *Character sum analogues of constant term identities for root systems*, Israel J. Math. **46** (1983), no. 3, 189–196, DOI 10.1007/BF02761951. MR733348

[31] R. J. Evans, *Hermite character sums*, Pacific J. Math. **122** (1986), no. 2, 357–390. MR831119

[32] R. J. Evans, *Character sums over finite fields*, Finite fields, coding theory, and advances in communications and computing (Las Vegas, NV, 1991), Lecture Notes in Pure and Appl. Math., vol. 141, Dekker, New York, 1993, pp. 57–73. MR1199822

[33] A. Erdélyi, W. Magnus, F. Oberhettinger, and F. G. Tricomi, *Higher transcendental functions. Vol. I*, Robert E. Krieger Publishing Co., Inc., Melbourne, Fla., 1981. Based on notes left by Harry Bateman; With a preface by Mina Rees; With a foreword by E. C. Watson; Reprint of the 1953 original. MR698779

[34] G. Faltings, *Endlichkeitssätze für abelsche Varietäten über Zahlkörpern* (German), Invent. Math. **73** (1983), no. 3, 349–366, DOI 10.1007/BF01388432. MR718935

[35] S. Frechette, K. Ono, and M. Papanikolas, *Gaussian hypergeometric functions and traces of Hecke operators*, Int. Math. Res. Not. **60** (2004), 3233–3262, DOI 10.1155/S1073792804132522. MR2096220

[36] S. Frechette, H. Swisher, and F.-T. Tu, *A cubic transformation formula for Appell-Lauricella hypergeometric functions over finite fields*, Res. Number Theory **4** (2018), no. 2, Paper No. 27, 27, DOI 10.1007/s40993-018-0119-9. MR3807415

[37] J. G. Fuselier, *Hypergeometric functions over $\mathbb{F}_p$ and relations to elliptic curves and modular forms*, Proc. Amer. Math. Soc. **138** (2010), no. 1, 109–123, DOI 10.1090/S0002-9939-09-10068-0. MR2550175

[38] I. Gessel and D. Stanton, *Strange evaluations of hypergeometric series*, SIAM J. Math. Anal. **13** (1982), no. 2, 295–308, DOI 10.1137/0513021. MR647127

[39] I. J. Good, *Generalizations to several variables of Lagrange's expansion, with applications to stochastic processes*, Proc. Cambridge Philos. Soc. **56** (1960), 367–380, DOI 10.1017/s0305004100034666. MR123021

[40] É. Goursat, *Sur l'équation différentielle linéaire, qui admet pour intégrale la série hypergéométrique* (French), Ann. Sci. École Norm. Sup. (2) **10** (1881), 3–142. MR1508709

[41] J. Greene, *Hypergeometric functions over finite fields*, Trans. Amer. Math. Soc. **301** (1987), no. 1, 77–101, DOI 10.2307/2000329. MR879564

[42] J. Greene, *Lagrange inversion over finite fields*, Pacific J. Math. **130** (1987), no. 2, 313–325. MR914104

[43] J. Greene, *Hypergeometric functions over finite fields and representations of* $SL(2,q)$, Rocky Mountain J. Math. **23** (1993), no. 2, 547–568, DOI 10.1216/rmjm/1181072576. MR1226188

[44] J. Greene and D. Stanton, *A character sum evaluation and Gaussian hypergeometric series*, J. Number Theory **23** (1986), no. 1, 136–148, DOI 10.1016/0022-314X(86)90009-0. MR840021

[45] A. Helversen-Pasotto, *L'identité de Barnes pour les corps finis* (French, with English summary), C. R. Acad. Sci. Paris Sér. A-B **286** (1978), no. 6, A297–A300. MR476707

[46] K. Ireland and M. Rosen, *A classical introduction to modern number theory*, 2nd ed., Graduate Texts in Mathematics, vol. 84, Springer-Verlag, New York, 1990, DOI 10.1007/978-1-4757-2103-4. MR1070716

[47] N. M. Katz, *Exponential sums and differential equations*, Annals of Mathematics Studies, vol. 124, Princeton University Press, Princeton, NJ, 1990, DOI 10.1515/9781400882434. MR1081536

[48] F. Klein, *Vorlesungen über die hypergeometrische Funktion* (German), Grundlehren der Mathematischen Wissenschaften [Fundamental Principles of Mathematical Sciences], vol. 39, Springer-Verlag, Berlin-New York, 1981. Reprint of the 1933 original. MR668700

[49] N. Koblitz, *The number of points on certain families of hypersurfaces over finite fields*, Compositio Math. **48** (1983), no. 1, 3–23. MR700577

[50] N. Koblitz, *p-adic numbers, p-adic analysis, and zeta-functions*, 2nd ed., Graduate Texts in Mathematics, vol. 58, Springer-Verlag, New York, 1984, DOI 10.1007/978-1-4612-1112-9. MR754003

[51] M. Koike, *Hypergeometric series over finite fields and Apéry numbers*, Hiroshima Math. J. **22** (1992), no. 3, 461–467. MR1194045

[52] M. Kontsevich and D. Zagier, *Periods*, Mathematics unlimited—2001 and beyond, Springer, Berlin, 2001, pp. 771–808. MR1852188

[53] E. E. Kummer, *Über die hypergeometrische Reihe . (Fortsetzung)* (German), J. Reine Angew. Math. **15** (1836), 127–172, DOI 10.1515/crll.1836.15.127. MR1578093

[54] J. L. Lagrange. Nouvelle méthode pour résoudre des équations littérales par le moyen de séries. *Mém. Acad. Roy. des Sci. et Belles-Lettres de Berlin*, 24, 1770.

[55] S. Lang, *Cyclotomic fields I and II*, 2nd ed., Graduate Texts in Mathematics, vol. 121, Springer-Verlag, New York, 1990. With an appendix by Karl Rubin, DOI 10.1007/978-1-4612-0987-4. MR1029028

[56] C. Lennon, *Gaussian hypergeometric evaluations of traces of Frobenius for elliptic curves*, Proc. Amer. Math. Soc. **139** (2011), no. 6, 1931–1938, DOI 10.1090/S0002-9939-2010-10609-3. MR2775369

[57] C. Lennon, *Trace formulas for Hecke operators, Gaussian hypergeometric functions, and the modularity of a threefold*, J. Number Theory **131** (2011), no. 12, 2320–2351, DOI 10.1016/j.jnt.2011.05.005. MR2832827

[58] W.-C. W. Li, *Barnes' identities and representations of* GL(2). *II. Non-Archimedean local field case*, J. Reine Angew. Math. **345** (1983), 69–92, DOI 10.1515/crll.1983.345.69. MR717887

[59] W.-C. W. Li, L. Long, and F.-T. Tu, *A Whipple $_7F_6$ formula revisited*, Matematica **1** (2022), no. 2, 480–530, DOI 10.1007/s44007-021-00015-6. MR4445932

[60] W.-C. W. Li and J. Soto-Andrade, *Barnes' identities and representations of* GL(2). *I. Finite field case*, J. Reine Angew. Math. **344** (1983), 171–179. MR716253

[61] Y.-H. Lin and F.-T. Tu, *Twisted Kloosterman sums*, J. Number Theory **147** (2015), 666–690, DOI 10.1016/j.jnt.2014.08.004. MR3276346

[62] The LMFDB Collaboration. *The L-functions and Modular Forms Database,* Home page of the Elliptic Curve 144.a3. http://www.lmfdb.org/EllipticCurve/Q/144/a/3, 2013. [Online; accessed 6 October 2015].

[63] The LMFDB Collaboration. *The L-functions and Modular Forms Database,* Home page of the Elliptic Curve 288.d3. http://www.lmfdb.org/EllipticCurve/Q/288/d/3, 2013. [Online; accessed 6 October 2015].

[64] The LMFDB Collaboration. *The L-functions and Modular Forms Database,* Home page of the Elliptic Curve 64.a3. http://www.lmfdb.org/EllipticCurve/Q/64/a/3, 2013. [Online; accessed 6 October 2015].

[65] L. Long, *On Shioda-Inose structures of one-parameter families of K3 surfaces*, J. Number Theory **109** (2004), no. 2, 299–318, DOI 10.1016/j.jnt.2004.06.009. MR2106484

[66] L. Long, *Hypergeometric evaluation identities and supercongruences*, Pacific J. Math. **249** (2011), no. 2, 405–418, DOI 10.2140/pjm.2011.249.405. MR2782677

[67] L. Long, *Some numeric hypergeometric supercongruences*, Vertex operator algebras, number theory and related topics, Contemp. Math., vol. 753, Amer. Math. Soc., [Providence], RI, [2020] ©2020, pp. 139–156, DOI 10.1090/conm/753/15169. MR4139242

[68] L. Long and R. Ramakrishna, *Some supercongruences occurring in truncated hypergeometric series*, Adv. Math. **290** (2016), 773–808, DOI 10.1016/j.aim.2015.11.043. MR3451938

[69] L. Long, F.-T. Tu, N. Yui, and W. Zudilin, *Supercongruences for rigid hypergeometric Calabi-Yau threefolds*, Adv. Math. **393** (2021), Paper No. 108058, 49, DOI 10.1016/j.aim.2021.108058. MR4330088

[70] D. McCarthy, *Transformations of well-poised hypergeometric functions over finite fields*, Finite Fields Appl. **18** (2012), no. 6, 1133–1147, DOI 10.1016/j.ffa.2012.08.007. MR3019189

[71] D. McCarthy, *Extending Gaussian hypergeometric series to the p-adic setting*, Int. J. Number Theory **8** (2012), no. 7, 1581–1612, DOI 10.1142/S1793042112500844. MR2968943

[72] D. McCarthy and M. A. Papanikolas, *A finite field hypergeometric function associated to eigenvalues of a Siegel eigenform*, Int. J. Number Theory **11** (2015), no. 8, 2431–2450, DOI 10.1142/S1793042115501134. MR3420754

[73] J. S. Milne, *Abelian Varieties* http://www.jmilne.org/math/CourseNotes/AV.pdf

[74] J. S. Milne, *Class Field Theory* http://www.jmilne.org/math/CourseNotes/CFT.pdf

[75] K. Ono, *Values of Gaussian hypergeometric series*, Trans. Amer. Math. Soc. **350** (1998), no. 3, 1205–1223, DOI 10.1090/S0002-9947-98-01887-X. MR1407498

[76] R. Osburn and C. Schneider, *Gaussian hypergeometric series and supercongruences*, Math. Comp. **78** (2009), no. 265, 275–292, DOI 10.1090/S0025-5718-08-02118-2. MR2448707

[77] R. Osburn and W. Zudilin, *On the (K.2) supercongruence of Van Hamme*, J. Math. Anal. Appl. **433** (2016), no. 1, 706–711, DOI 10.1016/j.jmaa.2015.08.009. MR3388817

[78] M. Petkova and H. Shiga, *A new interpretation of the Shimura curve with discriminant 6 in terms of Picard modular forms*, Arch. Math. (Basel) **96** (2011), no. 4, 335–348, DOI 10.1007/s00013-011-0235-4. MR2794089

[79] S. Ramanujan, *Modular equations and approximations to π [Quart. J. Math. **45** (1914), 350–372]*, Collected papers of Srinivasa Ramanujan, AMS Chelsea Publ., Providence, RI, 2000, pp. 23–39, DOI 10.1016/s0164-1212(00)00033-9. MR2280849

[80] D. P. Roberts and F. Rodriguez Villegas, *Hypergeometric motives*, Notices Amer. Math. Soc. **69** (2022), no. 6, 914–929, DOI 10.1090/noti2491. MR4442789

[81] J. Rouse, *Hypergeometric functions and elliptic curves*, Ramanujan J. **12** (2006), no. 2, 197–205, DOI 10.1007/s11139-006-0073-3. MR2286245

[82] A. Salerno, *Counting points over finite fields and hypergeometric functions*, Funct. Approx. Comment. Math. **49** (2013), no. 1, 137–157, DOI 10.7169/facm/2013.49.1.9. MR3127904

[83] H. A. Schwarz, *Ueber diejenigen Fälle, in welchen die Gaussische hypergeometrische Reihe eine algebraische Function ihres vierten Elementes darstellt* (German), J. Reine Angew. Math. **75** (1873), 292–335, DOI 10.1515/crll.1873.75.292. MR1579568

[84] J.-P. Serre, *Représentations linéaires des groupes finis* (French), Third revised edition, Hermann, Paris, 1978. MR543841

[85] J.-P. Serre, *Abelian l-adic representations and elliptic curves*, 2nd ed., Advanced Book Classics, Addison-Wesley Publishing Company, Advanced Book Program, Redwood City, CA, 1989. With the collaboration of Willem Kuyk and John Labute. MR1043865

[86] J. H. Silverman, *The arithmetic of elliptic curves*, Graduate Texts in Mathematics, vol. 106, Springer-Verlag, New York, 1986, DOI 10.1007/978-1-4757-1920-8. MR817210

[87] J. H. Silverman, *Advanced topics in the arithmetic of elliptic curves*, Graduate Texts in Mathematics, vol. 151, Springer-Verlag, New York, 1994, DOI 10.1007/978-1-4612-0851-8. MR1312368

[88] L. J. Slater, *Generalized hypergeometric functions*, Cambridge University Press, Cambridge, 1966. MR0201688

[89] J. Stienstra and F. Beukers, *On the Picard-Fuchs equation and the formal Brauer group of certain elliptic K3-surfaces*, Math. Ann. **271** (1985), no. 2, 269–304, DOI 10.1007/BF01455990. MR783555

[90] H. Swisher, *On the supercongruence conjectures of van Hamme*, Res. Math. Sci. **2** (2015), Art. 18, 21, DOI 10.1186/s40687-015-0037-6. MR3411813

[91] K. Takeuchi, *Arithmetic triangle groups*, J. Math. Soc. Japan **29** (1977), no. 1, 91–106, DOI 10.2969/jmsj/02910091. MR429744

[92] K. Takeuchi, *Commensurability classes of arithmetic triangle groups*, J. Fac. Sci. Univ. Tokyo Sect. IA Math. **24** (1977), no. 1, 201–212. MR463116

[93] F.-T. Tu and Y. Yang, *Evaluation of certain hypergeometric functions over finite fields*, SIGMA Symmetry Integrability Geom. Methods Appl. **14** (2018), Paper No. 050, 18, DOI 10.3842/SIGMA.2018.050. MR3803730

[94] F.-T. Tu and Y. Yang, *Algebraic transformations of hypergeometric functions and automorphic forms on Shimura curves*, Trans. Amer. Math. Soc. **365** (2013), no. 12, 6697–6729, DOI 10.1090/S0002-9947-2013-05960-0. MR3105767

[95] L. van Hamme, *Some conjectures concerning partial sums of generalized hypergeometric series*, p-adic functional analysis (Nijmegen, 1996), Lecture Notes in Pure and Appl. Math., vol. 192, Dekker, New York, 1997, pp. 223–236. MR1459212

[96] M. V. Vega, *Hypergeometric functions over finite fields and their relations to algebraic curves*, Int. J. Number Theory **7** (2011), no. 8, 2171–2195, DOI 10.1142/S1793042111004976. MR2873147

[97] R. Vidūnas, *Transformations of some Gauss hypergeometric functions*, J. Comput. Appl. Math. **178** (2005), no. 1-2, 473–487, DOI 10.1016/j.cam.2004.09.053. MR2127899

[98] R. Vidūnas, *Algebraic transformations of Gauss hypergeometric functions*, Funkcial. Ekvac. **52** (2009), no. 2, 139–180, DOI 10.1619/fesi.52.139. MR2547100

[99] M. Watkins, Hypergeometric motives notes, Preprint http://magma.maths.usyd.edu.au/~watkins/papers/known.pdf (2017).

[100] A. Weil, *Numbers of solutions of equations in finite fields*, Bull. Amer. Math. Soc. **55** (1949), 497–508, DOI 10.1090/S0002-9904-1949-09219-4. MR29393

[101] A. Weil, *Jacobi sums as "Grössencharaktere"*, Trans. Amer. Math. Soc. **73** (1952), 487–495, DOI 10.2307/1990804. MR51263

[102] F. J. W. Whipple, *On Well-Poised Series, Generalized Hypergeometric Series having Parameters in Pairs, each Pair with the Same Sum*, Proc. London Math. Soc. (2) **24** (1925), no. 4, 247–263, DOI 10.1112/plms/s2-24.1.247. MR1577160

[103] J. Wolfart, *Werte hypergeometrischer Funktionen* (German, with English summary), Invent. Math. **92** (1988), no. 1, 187–216, DOI 10.1007/BF01393999. MR931211

[104] K. Yamamoto, *On a conjecture of Hasse concerning multiplicative relations of Gaussian sums*, J. Combinatorial Theory **1** (1966), 476–489. MR213311

[105] M. Yoshida, *Fuchsian differential equations*, Aspects of Mathematics, E11, Friedr. Vieweg & Sohn, Braunschweig, 1987. With special emphasis on the Gauss-Schwarz theory, DOI 10.1007/978-3-663-14115-0. MR986252

[106] E. T. Whittaker and G. N. Watson, *A course of modern analysis*, Cambridge Mathematical Library, Cambridge University Press, Cambridge, 1996. An introduction to the general theory of infinite processes and of analytic functions; with an account of the principal transcendental functions; Reprint of the fourth (1927) edition, DOI 10.1017/CBO9780511608759. MR1424469

[107] M. Yoshida, *Hypergeometric functions, my love*, Aspects of Mathematics, E32, Friedr. Vieweg & Sohn, Braunschweig, 1997. Modular interpretations of configuration spaces, DOI 10.1007/978-3-322-90166-8. MR1453580

[108] W. Zudilin, *Ramanujan-type supercongruences*, J. Number Theory **129** (2009), no. 8, 1848–1857, DOI 10.1016/j.jnt.2009.01.013. MR2522708

Index

Nth power residue, 36
Nth symbol, 36
δ-function, 10
$\left(\dfrac{x}{\mathfrak{p}}\right)_m$: Nth power residue, 36
ε: trivial character, 10
$\iota_{\mathfrak{p}}(\cdot)$-function, 36
$\mathcal{J}_{\underline{a}}(\mathfrak{p})$: Weil's Jacobi sum, 38
ϕ: quadratic character, 10
$(*)$ condition, 1

algebraic hypergeometric series, 21
Andrews-Stanton cubic formula, 88
Artin L-function, 41

Bailey cubic transformation, 88
beta function, 7

characters
    quadratic character, ϕ, 10
    trivial character, ε, 10
Clausen formula, 55

decomposition group, 33
degenerate Bailey cubic
    transformation, 95

Euler reflection formula, 9
Euler transformation, 18

finite field analogues
    Andrews-Stanton cubic formula, 88
    Bailey cubic transformation, 89, 94
    binomial coefficient, 12
    Clausen formula, 55
    degenerate Bailey cubic
        transformation, 96
    Gauss evaluation formula, 49
    Gessel-Stanton cubic formula, 97

hypergeometric function $_{n+1}\mathbb{F}_n$, 29
Kummer evaluation formula, 48, 85
Kummer quadratic
    transformation, 77, 84
period function $_{n+1}\mathbb{P}_n$, 27, 38
Pfaff-Saalschütz evaluation
    formula, 65, 66, 87
primitive hypergeometric/period
    function, 30
rising factorial, 12
Frobenius conjugacy class, 33

Galois representation, 33
gamma function, 7
Gauss multiplication formula, 9
Gauss sum, 10
Gauss summation formula, 24
generalized Legendre curve, 42
Gessel-Stanton cubic formula, 97
Grössencharacter, 35

Hasse-Davenport relation, 11, 52
Helversen-Pasotto formula, 62
hypergeometric algebraic varieties, 29
hypergeometric differential equation, 16
hypergeometric functions $_{n+1}\mathbb{F}_n$ over
    finite fields, 29
hypergeometric functions $_{n+1}F_n$, 15

Jacobi sum, 10

Kummer evaluation theorem, 24
Kummer quadratic transformation, 24

Legendre curve, 43

monodromy group, 19
monodromy representation, 19
multiplicative character, 10

period functions $_{n+1}\mathbb{P}_n$, 38
period functions $_{n+1}\mathbb{P}_n$ over finite fields, 27
period functions $_{n+1}P_n$, 16
Pfaff transformation, 18
Pfaff-SaalSchütz evaluation formula, 24
Pfaff-Saalschütz evaluation formula
  in terms of $_3\mathbb{P}_2$, 65
  in terms of Jacobi sums, 66
  quadratic version, 87

primitive hypergeometric/period functions over finite fields, 30
projective monodromy group, 19, 21

Schwarz map, 17
Schwarz theorem, 17
Schwarz triangle $\Delta(p, q, r)$, 17, 21

triangle group, 20
  $(2, 2, n)$, 22
  $(2, 3, 3)$, 23
  (e_1, e_2, e_3), 21

unramified, 33

Yamamoto's example, 39

zeta function, 51

Editorial Information

To be published in the *Memoirs*, a paper must be correct, new, nontrivial, and significant. Further, it must be well written and of interest to a substantial number of mathematicians. Piecemeal results, such as an inconclusive step toward an unproved major theorem or a minor variation on a known result, are in general not acceptable for publication.

Papers appearing in *Memoirs* are generally at least 80 and not more than 200 published pages in length. Papers less than 80 or more than 200 published pages require the approval of the Managing Editor of the Transactions/Memoirs Editorial Board. Published pages are the same size as those generated in the style files provided for $\mathcal{AMS}$-LaTeX.

Information on the backlog for this journal can be found on the AMS website starting from `http://www.ams.org/memo`.

A Consent to Publish is required before we can begin processing your paper. After a paper is accepted for publication, the Providence office will send a Consent to Publish and Copyright Agreement to all authors of the paper. By submitting a paper to the *Memoirs*, authors certify that the results have not been submitted to nor are they under consideration for publication by another journal, conference proceedings, or similar publication.

Information for Authors

Memoirs is an author-prepared publication. Once formatted for print and on-line publication, articles will be published as is with the addition of AMS-prepared frontmatter and backmatter. Articles are not copyedited; however, confirmation copy will be sent to the authors.

Initial submission. The AMS uses Centralized Manuscript Processing for initial submissions. Authors should submit a PDF file using the Initial Manuscript Submission form found at `www.ams.org/submission/memo`, or send one copy of the manuscript to the following address: Centralized Manuscript Processing, MEMOIRS OF THE AMS, 201 Charles Street, Providence, RI 02904-2294 USA. If a paper copy is being forwarded to the AMS, indicate that it is for *Memoirs* and include the name of the corresponding author, contact information such as email address or mailing address, and the name of an appropriate Editor to review the paper (see the list of Editors below).

The paper must contain a *descriptive title* and an *abstract* that summarizes the article in language suitable for workers in the general field (algebra, analysis, etc.). The *descriptive title* should be short, but informative; useless or vague phrases such as "some remarks about" or "concerning" should be avoided. The *abstract* should be at least one complete sentence, and at most 300 words. Included with the footnotes to the paper should be the 2020 *Mathematics Subject Classification* representing the primary and secondary subjects of the article. The classifications are accessible from `www.ams.org/msc/`. The Mathematics Subject Classification footnote may be followed by a list of *key words and phrases* describing the subject matter of the article and taken from it. Journal abbreviations used in bibliographies are listed in the latest *Mathematical Reviews* annual index. The series abbreviations are also accessible from `www.ams.org/msnhtml/serials.pdf`. To help in preparing and verifying references, the AMS offers MR Lookup, a Reference Tool for Linking, at `www.ams.org/mrlookup/`.

Electronically prepared manuscripts. The AMS encourages electronically prepared manuscripts, with a strong preference for $\mathcal{AMS}$-LaTeX. To this end, the Society has prepared $\mathcal{AMS}$-LaTeX author packages for each AMS publication. Author packages include instructions for preparing electronic manuscripts, samples, and a style file that generates the particular design specifications of that publication series.

Authors may retrieve an author package for *Memoirs of the AMS* from `www.ams.org/journals/memo/memoauthorpac.html`. The *AMS Author Handbook* is available in PDF format from the author package link. The author package can also be obtained free of charge by sending email to `tech-support@ams.org` or from the Publication Division,

American Mathematical Society, 201 Charles St., Providence, RI 02904-2294, USA. When requesting an author package, please specify the publication in which your paper will appear. Please be sure to include your complete mailing address.

After acceptance. The source files for the final version of the electronic manuscript should be sent to the Providence office immediately after the paper has been accepted for publication. The author should also submit a PDF of the final version of the paper to the editor, who will forward a copy to the Providence office.

Accepted electronically prepared files can be submitted via the web at `www.ams.org/submit-book-journal/`, sent via FTP, or sent on CD to the Electronic Prepress Department, American Mathematical Society, 201 Charles Street, Providence, RI 02904-2294 USA. TeX source files and graphic files can be transferred over the Internet by FTP to the Internet node `ftp.ams.org` (130.44.1.100). When sending a manuscript electronically via CD, please be sure to include a message indicating that the paper is for the *Memoirs*.

Electronic graphics. Comprehensive instructions on preparing graphics are available at `www.ams.org/authors/journals.html`. A few of the major requirements are given here.

Submit files for graphics as EPS (Encapsulated PostScript) files. This includes graphics originated via a graphics application as well as scanned photographs or other computer-generated images. If this is not possible, TIFF files are acceptable as long as they can be opened in Adobe Photoshop or Illustrator.

Authors using graphics packages for the creation of electronic art should also avoid the use of any lines thinner than 0.5 points in width. Many graphics packages allow the user to specify a "hairline" for a very thin line. Hairlines often look acceptable when proofed on a typical laser printer. However, when produced on a high-resolution laser imagesetter, hairlines become nearly invisible and will be lost entirely in the final printing process.

Screens should be set to values between 15% and 85%. Screens which fall outside of this range are too light or too dark to print correctly. Variations of screens within a graphic should be no less than 10%.

Any graphics created in color will be rendered in grayscale for the printed version unless color printing is authorized by the Managing Editor and the Publisher. In general, color graphics will appear in color in the online version.

Inquiries. Any inquiries concerning a paper that has been accepted for publication should be sent to `memo-query@ams.org` or directly to the Electronic Prepress Department, American Mathematical Society, 201 Charles St., Providence, RI 02904-2294 USA.

Editors

This journal is designed particularly for long research papers, normally at least 80 pages in length, and groups of cognate papers in pure and applied mathematics. Papers intended for publication in the *Memoirs* should be addressed to one of the following editors. The AMS uses Centralized Manuscript Processing for initial submissions to AMS journals. Authors should follow instructions listed on the Initial Submission page found at www.ams.org/memo/memosubmit.html.

Managing Editor: Henri Darmon, Department of Mathematics, McGill University, Montreal, Quebec H3A 0G4, Canada; e-mail: darmon@math.mcgill.ca

1. GEOMETRY, TOPOLOGY & LOGIC

    Coordinating Editor: Richard Canary, Department of Mathematics, University of Michigan, Ann Arbor, MI 48109-1043 USA; e-mail: canary@umich.edu

    Algebraic topology, Michael Hill, Department of Mathematics, University of California Los Angeles, Los Angeles, CA 90095 USA; e-mail: mikehill@math.ucla.edu

    Logic, Mariya Ivanova Soskova, Department of Mathematics, University of Wisconsin–Madison, Madison, WI 53706 USA; e-mail: msoskova@math.wisc.edu

    Low-dimensional topology and geometric structures, Richard Canary

    Symplectic geometry, Yael Karshon, School of Mathematical Sciences, Tel-Aviv University, Tel Aviv, Israel; and Department of Mathematics, University of Toronto, Toronto, Ontario M5S 2E4, Canada; e-mail: karshon@math.toronto.edu

2. ALGEBRA AND NUMBER THEORY

    Coordinating Editor: Henri Darmon, Department of Mathematics, McGill University, Montreal, Quebec H3A 0G4, Canada; e-mail: darmon@math.mcgill.ca

    Algebra, Radha Kessar, Department of Mathematics, City, University of London, London EC1V 0HB, United Kingdom; e-mail: radha.kessar.1@city.ac.uk

    Algebraic geometry, Lucia Caporaso, Department of Mathematics and Physics, Roma Tre University, Largo San Leonardo Murialdo, I-00146 Rome, Italy; e-mail: LCedit@mat.uniroma3.it

    Analytic number theory, Lillian B. Pierce, Department of Mathematics, Duke University, 120 Science Drive Box 90320, Durham, NC 27708 USA; e-mail: pierce@math.duke.edu

    Arithmetic geometry, Ted C. Chinburg, Department of Mathematics, University of Pennsylvania, Philadelphia, PA 19104-6395 USA; e-mail: ted@math.upenn.edu

    Commutative algebra, Irena Peeva, Department of Mathematics, Cornell University, Ithaca, NY 14853 USA; e-mail: irena@math.cornell.edu

    Number theory, Henri Darmon

3. GEOMETRIC ANALYSIS & PDE

    Coordinating Editor: Alexander A. Kiselev, Department of Mathematics, Duke University, 120 Science Drive, Rm 117 Physics Bldg, Durham, NC 27708 USA; e-mail: kiselev@math.duke.edu

    Differential geometry and geometric analysis, Ailana M. Fraser, Department of Mathematics, University of British Columbia, 1984 Mathematics Road, Room 121, Vancouver BC V6T 1Z2, Canada; e-mail: afraser@math.ubc.ca

    Harmonic analysis and partial differential equations, Monica Visan, Department of Mathematics, University of California Los Angeles, 520 Portola Plaza, Los Angeles, CA 90095 USA; e-mail: visan@math.ucla.edu

    Partial differential equations and functional analysis, Alexander A. Kiselev

    Real analysis and partial differential equations, Joachim Krieger, Bâtiment de Mathématiques, École Polytechnique Fédérale de Lausanne, Station 8, 1015 Lausanne Vaud, Switzerland; e-mail: joachim.krieger@epfl.ch

4. ERGODIC THEORY, DYNAMICAL SYSTEMS & COMBINATORICS

    Coordinating Editor: Vitaly Bergelson, Department of Mathematics, Ohio State University, 231 W. 18th Avenue, Columbus, OH 43210 USA; e-mail: vitaly@math.ohio-state.edu

    Algebraic and enumerative combinatorics, Jim Haglund, Department of Mathematics, University of Pennsylvania, Philadelphia, PA 19104 USA; e-mail: jhaglund@math.upenn.edu

    Probability theory, Robin Pemantle, Department of Mathematics, University of Pennsylvania, 209 S. 33rd Street, Philadelphia, PA 19104 USA; e-mail: pemantle@math.upenn.edu

    Dynamical systems and ergodic theory, Ian Melbourne, Mathematics Institute, University of Warwick, Coventry CV4 7AL, United Kingdom; e-mail: I.Melbourne@warwick.ac.uk

    Ergodic theory and combinatorics, Vitaly Bergelson

5. ANALYSIS, LIE THEORY & PROBABILITY

    Coordinating Editor: Stefaan Vaes, Department of Mathematics, Katholieke Universiteit Leuven, Celestijnenlaan 200B, B-3001 Leuven, Belgium; e-mail: stefaan.vaes@wis.kuleuven.be

    Functional analysis and operator algebras, Stefaan Vaes

    Harmonic analysis, PDEs, and geometric measure theory, Svitlana Mayboroda, School of Mathematics, University of Minnesota, 206 Church Street SE, 127 Vincent Hall, Minneapolis, MN 55455 USA; e-mail: svitlana@math.umn.edu

    Probability theory and stochastic analysis, Davar Khoshnevisan, Department of Mathematics, The University of Utah, Salt Lake City, UT 84112 USA; e-mail: davar@math.utah.edu

SELECTED PUBLISHED TITLES IN THIS SERIES

1377 **Jacob Bedrossian, Pierre Germain, and Nader Masmoudi,** Dynamics Near the Subcritical Transition of the 3D Couette Flow II: Above Threshold Case, 2022

1376 **Lucia Di Vizio, Charlotte Hardouin, and Anne Granier,** Intrinsic Approach to Galois Theory of q-Difference Equations, 2022

1375 **Cai Heng Li and Binzhou Xia,** Factorizations of Almost Simple Groups with a Solvable Factor, and Cayley Graphs of Solvable Groups, 2022

1374 **Jan Kohlhaase,** Coefficient Systems on the Bruhat-Tits Building and Pro-p Iwahori-Hecke Modules, 2022

1373 **Yongsheng Han, Ming-Yi Lee, Ji Li, and Brett Wick,** Maximal Functions, Littlewood–Paley Theory, Riesz Transforms and Atomic Decomposition in the Multi-Parameter Flag Setting, 2022

1372 **François Charest and Chris Woodward,** Floer Cohomology and Flips, 2022

1371 **H. Flenner, S. Kaliman, and M. Zaidenberg,** Cancellation for surfaces revisited, 2022

1370 **Michele D'Adderio, Alessandro Iraci, and Anna Vanden Wyngaerd,** Decorated Dyck Paths, Polyominoes, and the Delta Conjecture, 2022

1369 **Stefano Burzio and Joachim Krieger,** Type II blow up solutions with optimal stability properties for the critical focussing nonlinear wave equation on $\mathbb{R}^{3+1}$, 2022

1368 **Dounnu Sasaki,** Subset currents on surfaces, 2022

1367 **Mark Gross, Paul Hacking, and Bernd Siebert,** Theta Functions on Varieties with Effective Anti-Canonical Class, 2022

1366 **Miki Hirano, Taku Ishii, and Tadashi Miyazaki,** Archimedean Zeta Integrals for $GL(3) \times GL(2)$, 2022

1365 **Alessandro Andretta and Luca Motto Ros,** Souslin Quasi-Orders and Bi-Embeddability of Uncountable Structures, 2022

1364 **Marco De Renzi,** Non-Semisimple Extended Topological Quantum Field Theories, 2022

1363 **Alan Hammond,** Brownian Regularity for the Airy Line Ensemble, and Multi-Polymer Watermelons in Brownian Last Passage Percolation, 2022

1362 **John Voight and David Zureick-Brown,** The Canonical Ring of a Stacky Curve, 2022

1361 **Nuno Freitas and Alain Kraus,** On the Symplectic Type of Isomorphisms of the p-Torsion of Elliptic Curves, 2022

1360 **Alexander V. Kolesnikov and Emanuel Milman,** Local L^p-Brunn-Minkowski Inequalities for $p < 1$, 2022

1359 **Franck Barthe and Paweł Wolff,** Positive Gaussian Kernels Also Have Gaussian Minimizers, 2022

1358 **Swee Hong Chan and Lionel Levine,** Abelian Networks IV. Dynamics of Nonhalting Networks, 2022

1357 **Camille Laurent and Matthieu Léautaud,** Tunneling Estimates and Approximate Controllability for Hypoelliptic Equations, 2022

1356 **Matthias Grüninger,** Cubic Action of a Rank One Group, 2022

1355 **David A. Craven,** Maximal PSL_2 Subgroups of Exceptional Groups of Lie Type, 2022

1354 **Gian Paolo Leonardi, Manuel Ritoré, and Efstratios Vernadakis,** Isoperimetric Inequalities in Unbounded Convex Bodies, 2022

1353 **Clifton Cunningham, Andrew Fiori, Ahmed Moussaoui, James Mracek, and Bin Xu,** Arthur Packets for p-adic Groups by Way of Microlocal Vanishing Cycles of Perverse Sheaves, with Examples, 2022

For a complete list of titles in this series, visit the
AMS Bookstore at **www.ams.org/bookstore/memoseries/**.